FLOWING WATER
FISH CULTURE

FLOWING WATER FISH CULTURE

Richard W. Soderberg
Fisheries Program
Mansfield University
Mansfield, Pennsylvania

CRC Press
Taylor & Francis Group
Boca Raton   London   New York

CRC Press is an imprint of the
Taylor & Francis Group, an **informa** business

First published 1995 by Lewis Publishers, Inc.

Published 2019 by CRC Press
Taylor & Francis Group
6000 Broken Sound Parkway NW, Suite 300
Boca Raton, FL 33487-2742

© 1995 by Taylor & Francis Group, LLC
CRC Press is an imprint of Taylor & Francis Group, an Informa business

First issued in paperback 2019

No claim to original U.S. Government works

ISBN-13: 978-0-367-44933-9 (pbk)
ISBN-13: 978-1-56670-081-8 (hbk)

Visit the Taylor & Francis Web site at
http://www.taylorandfrancis.com

and the CRC Press Web site at
http://www.crcpress.com

Library of Congress Card Number 94-18433

Library of Congress Cataloging-in-Publication Data

Soderberg, Richard W.
  Flowing water fish culture / Richard W. Soderberg
      p.   cm.
    Inclu
    ISBN
    1. Fish culture—Water supply.   I. Title.
  SH154.S64   1994
  693.3'13—dc20
                                                            94-18433
                                                               CIP

PREFACE

I began this project when I was called upon to teach a series of courses at Mansfield University on fish husbandry technology. Finding no suitable text on intensive aquaculture methods, I began to develop my own. The present volume is a status report on that continuing process. As chapters were completed and sent to colleagues for review, it soon became evident that the book would have a wider utility than as a college text. Aquaculture is expanding worldwide at a phenomenal rate. Much of this expansion is in the development of intensive systems and the intensification of existing ones. *Flowing Water Fish Culture* should be a useful reference for those involved in the development of intensive aquacultures.

The reader will soon observe that this is not a book on how to grow fish; quite satisfactory texts on the process of fish culture are available. Rather, it is an account of the science of fish husbandry in a stream of water and is intended to convey that science in a manner appropriate for use by university students and teachers, and others involved in fish production research and development.

A comment on units of measure is in order. Both the English and metric systems are used by American fish culturists. The principle American journal of aquaculture publishes most of its papers in the English system of measure while others use the metric system. Some state fishery agencies have adopted the metric system while others use the English system. Thus the student and practitioner of aquaculture technology must be able to use both systems interchangeably. The problem sets at the end of each chapter provide ample opportunity to develop that skill.

Completion of this book would not have been possible without constructive criticism from my colleagues and students at Auburn University and Mansfield University. I would especially like to thank Professor Claude E. Boyd for so eloquently demonstrating that the solution to biological problems lies in the quantitative investigation of their components.

Richard W. Soderberg

THE AUTHOR

Richard W. Soderberg is a professor and director of the Fisheries Program at Mansfield University of Pennsylvania. He received his B.S. degree from the University of Wisconsin at Madison and his M.S. and Ph.D. degrees from Auburn University, Auburn, Alabama.

Dr. Soderberg has published extensively in the areas of fish culture, aquaculture system design, water quality and fish health. His research has included various aspects of the husbandry of yellow perch, lake trout, Atlantic salmon, tilapia and walleye. Additionally, he has studied the management of sport fish populations in farm ponds, blood chemistry of fish exposed to acid pollution, and angler use and harvest surveys. He has presented numerous papers on these topics at national and international conferences, workshops and symposia. In addition, he has served as an associate editor of the *Progressive Fish-Culturist* and is an appointed member of the Pennsylvania Fish and Boat Commission's planning task force.

Dr. Soderberg is active in the American Fisheries Society where he has served two terms as Pennsylvania Chapter President and was recently awarded for outstanding service to the Society by the Pennsylvania chapter. He is a member of the American Institute of Fishery Research Biologists where he was promoted to Fellow in 1994.

TABLE OF CONTENTS

Dedicated to
Arlette M. Soderberg
for the inspiration, encouragement
and example for a scholarly life

CHAPTER 1

Flowing Water Fish Culture

The practice of flowing water fish culture in the U.S. began in the mid-1800s in response to dwindling numbers of the native brook trout. The facilities were little more than fenced-off portions of streams resembling trout habitat. Eventually, all states with trout fisheries established fishery agencies and built hatcheries and the artificial reproduction and husbandry of salmonids was documented.

There are two basic considerations for the controlled growth of fish in a stream of water. First, the medium that supplies oxygen for respiration and flushes away metabolic waste is water, which has a low affinity for oxygen and occurs in finite quantities. This problem becomes obvious when aquaculture is compared to terrestrial animal husbandry in air, which contains a great deal of oxygen and is available in unlimited supply. Secondly, fish are cold-blooded and grow satisfactorily within rather narrow temperature ranges. Thus, water temperature determines which species, if any, may be produced. Warm-blooded animals, on the other hand, use food energy to maintain their optimum growth temperatures regardless of environmental temperature.

In the 1950s David Haskell, an engineer employed by the New York Conservation Department, first applied analytical investigation to the art of flowing water fish culture. Haskell's quantitative approach to the definition of chemical and biological parameters affecting fish in confinement allowed fish culture to progress from an art to a science.

Haskell's pioneering work resulted in the elucidation of five basic principles upon which our present understanding of flowing water fish culture is based:

1. At constant temperature, fish growth, in units of length, is linear over time until sexual maturity is approached.
2. The growth rate of fish, in units of length, is proportional to temperature. Therefore, if the growth rate at one temperature is known, the growth rate at another temperature may be predicted.
3. Feeding rates can be rationally calculated based on estimated food conversion, metabolic characteristics, and the anticipated growth rate.

4. The maximum permissible weight of fish that can be supported in a rearing unit is determined by the depletion of oxygen and the accumulation of metabolic wastes.
5. Oxygen consumption and metabolite production progress in proportion to the amount of food fed.

Based upon this framework, flowing water fish culture has become a quantitative agricultural science. The technology of that science is the subject of the chapters that follow.

CHAPTER 2

Fish Growth In Hatcheries

GROWTH RATES OF CULTURED FISH IN RELATION TO TEMPERATURE

Fish Growth Models based on Temperature Units

Haskell (1959) plotted trout growth rate against temperature in the range of 42 to 52°F, and found that they were linearly related with an intercept at 38.6°F (Figure 2.1). Growth does not cease at 38.6°F, but the intercept is necessary to establish the line from which growth is predicted. Haskell (1959) defined a temperature unit (TU) as 1°F over 38.6°F for 1 month. For example, if the monthly water temperature averaged 50°F, 50 − 38.6 = 11.4 TU would be available during that month. Haskell reported that approximately 21 TU were required per inch of trout growth in New York State hatcheries. In an experiment where brook trout were grown under conditions of "optimum care", 16.2 TU were required per inch of growth (Haskell 1959), but Haskell doubted that trout could grow that fast under hatchery conditions. Modern hatcheries experience growth rates of approximately 17 TU/in., possibly due to improvements in hatchery management that have occurred since Haskell's work. The actual growth rate, in temperature units required per length increment of growth, should be determined for each hatchery, species, and strain of fish.

The U.S. Fish and Wildlife Service (Piper et al. 1982) has adopted monthly temperature units (MTU), defined as the average monthly water temperature minus the freezing point of water. Therefore the MTU theory differs from Haskell's TU model by assuming linear growth at temperatures cooler than 42°F. This procedure forces the growth line through the origin, which could result in a substantial error in growth rate estimation, especially for warmwater fish whose zero-growth intercepts are at higher temperatures than for salmonids. In Figure 2.2, Haskell's (1959) data for brook trout are shown on a growth vs. temperature plot containing a regression line of best fit for the data when it is forced through the origin of the graph. It is evident that ignoring the zero-growth intercept compromises the accuracy of the model.

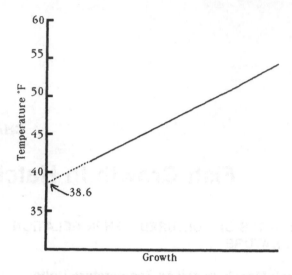

Figure 2.1. Plot of brook trout growth in units of fish length at temperatures from 42 to 52°F. (Haskell 1959).

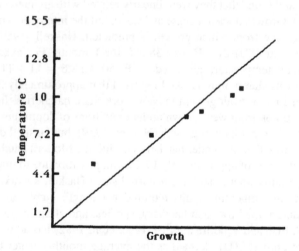

Figure 2.2. When Haskell's (1959) data are fitted to a regression line forced through the origin of the graph in accordance with the MTU growth prediction method (Piper et al. 1982), the accuracy of the model is reduced.

Research at the U.S. Fish and Wildlife Service Fish Culture Development Center in Bozeman, Montana has defined the MTU requirements for some salmonid species. Brook trout (Dwyer et al. 1981a) required 7.2 Centigrade MTU/cm of growth in the temperature range of 7 to 19°C, but only 6.1 MTU/cm from 10 to 16°C (Table 2.1). Rainbow trout (Dwyer et al. 1981b) required 5.8 MTU/cm at 13°C and an average of 7.1 MTU/cm in the temperature

Table 2.1. MTU (average monthly temperature, °C)
required per Centimeter of Brook Trout Growth
at Temperatures from 4 to 19°C

Exposure temperature, °C	MTU required/ cm of growth
4	15.2a
7	8.7 b
10	5.8 bc
13	5.7 bc
16	6.8 bc
19	9.1 b
Mean (7–19°C)	7.2
Mean (10–16°C)	6.1

Note: Values followed by the same letter are not significantly
different (P > 0.05).
After Dwyer et al. (1981a).

range of 7 to 19°C (Table 2.2). In the temperature range of 4 to 16°C, lake trout
(Dwyer et al. 1981c) required 5.4 Centigrade MTU/cm of growth (Table 2.3).
Steelhead trout (Dwyer et al. 1982) required 10.9 MTU/cm in the range from
4 to 19°C, but only 8.5 MTU/cm at 13°C (Table 2.4). An average of 10.1
MTU/cm were required for Atlantic salmon (Dwyer and Piper 1987) at temperatures ranging from 7 to 16°C, but growth efficiency was significantly
reduced at 4 and 19°C (Table 2.5).

Notice that the temperature units required per centimeter of growth for these
species are generally greater at the extremes of the temperature ranges tested
than within a narrower, optimum temperature range. This demonstrates that
fish grow most efficiently within rather narrow ranges of temperature.

Andrews et al. (1972) presented data on the growth of channel catfish at
temperatures from 24 to 30°C. When their reported final weights are converted
to length (Piper et al. 1982), and length increment is plotted against temperature, the intercept is negative. Thus, the TU procedure is inappropriate for these
data. When the MTU method of growth projection is applied, a useable growth
model for catfish is obtained (Table 2.6). An average of 6.3 Centigrade MTU
were required per centimeter of catfish growth at temperatures from 24 to
30°C.

Soderberg (1990) found that the growth rate vs. temperature plot for blue
tilapia, *Oreochromis aureus,* had an intercept of 17.8°C (Figure 2.3). An
average of 6.9 Centigrade TU were required per centimeter of tilapia growth
in the temperature range of 20 to 30°C, where the temperature units available
per month equal the average monthly water temperature minus 17.8°C (Table
2.7).

Meade et al. (1983) studied the temperature related growth of tiger muskellunge, *Esox lucius × E. masquinongy.* In the temperature range of 14 to 24°C,
3 to 4 cm early fingerlings required 3.8 Centigrade MTU/cm of growth. Larger
fish (12 to 13 cm) required 5.8 MTU/cm from 18 to 28°C, and 5.1 MTU/cm
from 18 to 24°C (Table 2.8).

Table 2.2. MTU (average monthly temperature, °C) required per Centimeter of Rainbow Trout Growth at Temperatures from 4 to 19°C

Exposure temperature, °C	MTU required/ cm of growth
4	12.4a
7	8.7 b d
10	6.6 bcd
13	5.8 b
16	6.3 bcd
19	8.2 bcd
Mean (7–19°C)	7.1

Note: Values followed by the same letter are not significantly different ($p > 0.05$).
After Dwyer et al. (1981b).

Table 2.3. MTU (average monthly temperature, °C) Required per Centimeter of Lake Trout Growth at Temperatures from 4 to 16°C

Exposure temperature, °C	MTU required/ cm of growth
4	5.0a
7	4.5a
10	5.1a
13	5.4a
16	7.0a
Mean	5.4

Note: Values followed by the same letter are not significantly different ($p > 0.05$).
After Dwyer et al. (1981c).

Table 2.4. MTU (average monthly temperature, °C) Required per Centimeter of Steelhead Trout Growth at Temperatures from 4 to 19°C

Exposure temperature, °C	MTU required/ cm of growth
4	15.0a
7	9.6a
10	9.5a
13	8.5b
16	10.0a
19	12.7a
Mean	10.9

Note: Values followed by the same letter are not significantly different ($p > 0.05$).
After Dwyer et al. (1982).

Table 2.5. MTU (average monthly temperature, °C)
           Required per Centimeter of Atlantic Salmon
           Growth at Temperatures from 4 to 19°C

Exposure temperature, °C	MTU required/ cm of growth
4	14.6a
7	10.9b
10	10.0b
13	9.9b
16	9.7b
19	12.4a
Mean (7–16°C)	10.1

Note: Values followed by the same letter are not significantly
       different (p > 0.05)
From Dwyer, W. P. and Piper, R. G., Progressive Fish-Culturist,
49, 58, 1987. With permission.

Table 2.6. MTU (average monthly temperature, C)
           Required per Centimeter of Channel Catfish
           Growth at Temperatures from 24 to 30°C

Exposure temperature, °C	MTU required/ cm of growth
24	6.1
26	6.1
28	6.3
30	6.8
Mean	6.3

Modified after Andrews et al. (1972).

The temperature unit and monthly temperature requirements of cultured fish
are summarized in Table 2.9.

Regression Models for Fish Growth

Because of the confusion associated with different temperature unit models
with different intercepts and units of measure, Soderberg (1990) proposed that
the independent variable, temperature, be regressed against growth in accordance with the general linear model. The ΔL value required for fish growth
projection can then be calculated from the regression equation $Y = a + bX$,
where $Y = \Delta L$, a = the intercept, b = the slope, and X = the fixed variable,
temperature. The reversed plot of Soderberg's (1990) tilapia growth data is
shown in Figure 2.4. Soderberg (1992) subjected the available fish growth data
to regression analysis in order to develop linear growth models for brook trout,
rainbow trout, lake trout, steelhead, Atlantic salmon, channel catfish, tiger
muskellunge, and blue tilapia.

Brook trout: Growth at 4°C was significantly less efficient than at temperatures from 7 to 19°C (Dwyer et al. 1981a). The growth equation for the

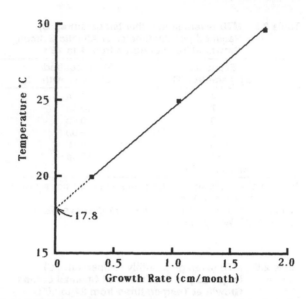

Figure 2.3. Growth of blue tilapia, *Oreochromis aureas,* in cm/month at temperatures from 20 to 30°C. (From Soderberg, R. W., *Progressive Fish-Culturist,* 52, 157, 1990. With permission.)

Table 2.7. **TU (1 TU = 1°C greater than 17.8°C for one month) Required per Centimeter of Blue Tilapia Growth at Temperatures from 20 to 30°C**

Exposure temperature, °C	TU required/ cm of growth
20	6.7
25	7.0
30	6.9
Mean	6.9

From Soderberg, R. W., *Progressive Fish-Culturist,* 52, 156, 1990. With permission.

temperature range of 4 to 19°C is $\Delta L = 0.155 + 0.0355\ T$ (ΔL is in mm/day and T is temperature in °C for all equations presented here) with a coefficient of determination (r^2) of 0.637. The r^2 value is improved to 0.792 in the more efficient temperature range of 7 to 19°C. The growth equation over this temperature range is $\Delta L = 0.006 + 0.0455\ T$. Although temperature unit requirements did not differ significantly over the temperature range of 7 to 19°C, actual growth was less at 19°C than at 16°C (Dwyer et al. 1981a). Thus, the linear growth model is further improved by narrowing the temperature range to 7 to 16°C, where $\Delta L = -0.068 + 0.0578\ T$ ($r^2 = 0.882$). The values derived from Haskell's (1959) data are better correlated ($r^2 > 0.99$) than those of Dwyer et al. (1981a), probably due to the narrower temperature range (5.5 to 12.2°C) of his determinations. The equation resulting from regression analysis of Haskell's data is $\Delta L = 0.348 + 0.0944\ T$ and gives ΔL values similar to

Table 2.8. MTU (average monthly temperature, °C) Required
per Centimeter of Tiger Muskellunge Early and
Advanced Fingerling Growth at Temperatures from
14 to 28°C

Exposure temperature, °C	MTU requirements/cm of growth	
	Early fingerlings (3–4 cm)	Advanced fingerlings (12–13 cm)
14	3.9	—
16	3.8	—
18	3.8	5.0
19	3.7	—
20	3.7	4.8
21	3.9	5.0
22	3.7	5.2
23	—	5.1
24	3.8	5.3
26	—	6.7
28	—	9.3
Mean	3.8	5.8

After Meade et al. (1983).

Table 2.9. Summary of Temperature Unit or Monthly Temperature Unit
Requirements of Cultured Fish

Fish species	TU or MTU/cm	Temperature range, °C	Ref.
Brook Trout	16.2[1]	6–11	Haskell 1959
Brook Trout	7.2[2]	7–19	Dwyer et al. 1981a
Brook Trout	6.1[2]	10–16	Dwyer et al. 1981a
Rainbow Trout	7.1[2]	7–19	Dwyer et al. 1981b
Lake Trout	5.4[2]	4–16	Dwyer et al. 1981c
Steelhead Trout	10.9[2]	4–19	Dwyer et al. 1982
Steelhead Trout	8.5[2]	13	Dwyer et al. 1982
Atlantic Salmon	10.1[2]	7–16	Dwyer and Piper 1987
Channel Catfish	6.3[2]	24–30	Andrews et al. 1972
Blue Tilapia	6.9[3]	20–30	Soderberg 1990
Tiger Muskellunge			
3–4 cm	3.8[2]	14–24	Meade et al. 1983
12–13 cm	5.8[2]	18–28	Meade et al. 1983
12–13 cm	5.1[2]	18–24	Meade et al. 1983

[1] TU = Mean monthly temperature in °F − 38.6°F
[2] MTU = Mean monthly temperature in °C − 0°C
[3] TU = Mean monthly temperature in °C − 17.8°C

those predicted by the equation derived from Dwyer's data in the temperature
range of 7 to 16°C.

Rainbow trout: The MTU requirements were significantly greater at 4°C
than at other test temperatures up to 19°C. Growth efficiency did not differ
significantly in the temperature range of 7 to 19°C, but actual growth at 19°C
was less than that at 16°C (Dwyer et al. 1981b). The growth equations for the
temperature ranges of 4 to 19, 7 to 19 and 7 to 16°C are $\Delta L = -0.040 + 0.505$
T ($r^2 = 0.886$), $\Delta L = 0.043 + 0.0450 T$ ($r^2 = 0.801$) and $\Delta L = -0.167 + 0.066$
T ($r^2 = 0.971$), respectively.

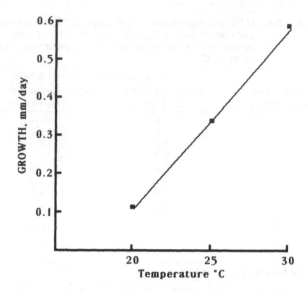

Figure 2.4. Plot of Soderberg's (1990) tilapia growth data with the axes reversed to allow the calculation of the daily estimated length increment (ΔL) from the general linear model. ΔL = $-0.853 + 0.048$ T ($r^2 > 0.99$) in the temperature range of 20 to 30°C.

Lake trout: The MTU requirements for growth did not significantly differ over the temperature range of 4 to 16°C (Dwyer et al. 1981c), but fish at 16°C grew less than those at 13°C. The linear growth equation for the temperature range of 4 to 16°C is ΔL = $0.176 + 0.0426$ T ($r^2 = 0.858$) and the equation for the temperature range of 4 to 13°C is ΔL = $0.0622 + 0.588$ T ($r^2 = 0.979$).

Steelhead: The MTU requirements per centimeter of growth was significantly lower at 13°C than at 4, 7, 10, 16, or 19°C (Dwyer et al. 1982). The regression equation over the temperature range of 4 to 19°C is ΔL = $0.0329 + 0.0294$ T ($r^2 = 0.856$). The growth equation is ΔL = $0.0148 + 0.0343$ T ($r^2 = 0.937$) over the temperature range of 7 to 16°C.

Atlantic salmon: Fish grown at 4 and 19°C required significantly more MTU per unit of length increase than those held at 7, 10, 13 and 16°C. Growth at 19°C was less than that at 16°C (Dwyer and Piper 1987). The growth equation over the entire range of temperatures tested is ΔL = $0.0043 + 0.0306$ T ($r^2 = 0.926$). The improved regression over the temperature range of 7 to 16°C is ΔL = $-0.0429 + 0.0371$ T ($r^2 > 0.99$).

Channel catfish: Andrews et al. (1973) studied catfish growth at temperatures from 24 to 30°C. Calculated daily incremental growth increased with temperature from 24 to 28°C, but was less at 30 than at 28°C. The regression equation for catfish growth is ΔL = $0.612 + 0.0298$ T ($r^2 = 0.825$) at temperatures from 24 to 30°C and ΔL = $0.195 + 0.0463$ T ($r^2 = 0.991$) from 24 to 28°C.

Tiger muskellunge: The growth of early fingerlings (3 to 4 cm) was well correlated with temperature in the range of 14 to 24°C (Meade et al. 1983). The regression equation calculated from their data is $\Delta L = -0.0548 + 0.912\ T$ ($r^2 = 0.985$). The growth rate of advanced fingerlings (12 to 13 cm) was greatest at 23 and 24°C, but declined at higher and lower temperatures (Meade et al. 1983). Thus, temperature and daily incremental growth for larger tiger muskellunge were not correlated ($r^2 = 0.12$) over the entire tested temperature range of 18 to 28°C. The growth equation for the temperature range 18 to 24°C is $\Delta L = 0.394 + 0.0471\ T$ ($r^2 = 0.864$).

Blue tilapia: Soderberg (1990) reported that the regression of daily growth increment on temperature in the range of 20 to 30°C is $\Delta L = -0.853 + 0.048\ T$ ($r^2 > 0.99$).

The regression equations described above and their coefficients of determination are summarized in Table 2.10.

PROJECTION OF FISH GROWTH IN TIME

Temperature Unit Procedure

The ability to project the size of fish in advance is necessary for determining daily feed rates, feed orders, egg orders, stocking dates, and production schedules. The following example illustrates how temperature unit methods are used to predict fish growth. Consider a lot of brook trout being reared at a hatchery where the average monthly water temperature is predicted to be 8.2°C. If 7.2 MTU are required per centimeter of growth, the expected monthly gain in length is

$$\frac{8.2\ \text{MTU}}{\text{month}} \times \frac{\text{cm}}{7.2\ \text{MTU}} = \frac{1.14\ \text{cm}}{\text{month}}$$

Similarly, using TU, if the average monthly water temperature is 49°F, 49 − 38.6, or 10.4 TU are available for that month. If 17 TU are required per inch of growth, the monthly length gain is

$$\frac{10.4\ \text{TU}}{\text{month}} \times \frac{\text{in.}}{17\ \text{TU}} = \frac{0.61\ \text{in.}}{\text{month}}$$

Fish growth is projected in units of length, but the weight of fish of known length is often required for fish hatchery management. Weight is easily calculated from length with the following expression:

$$W = KL^3 \qquad\qquad 2.1$$

where W = fish weight, L = fish length, and K = condition factor. The condition factor is species- and strain-dependent, but remains constant through the linear phase of growth if diet and health are adequate.

Table 2.10 Linear Models for Fish Growth in Hatcheries

Species	Growth model	Temperature range, °C	r^2	Ref.
Brook Trout	$\Delta L = -0.348 + 0.0944\ T$	5.5–12.2	>0.99	Haskell 1959
Brook Trout	$\Delta L = 0.155 + 0.0355\ T$	4–19	0.637	Dwyer et al. 1981a
Brook Trout	$\Delta L = 0.006 + 0.0455\ T$	7–19	0.792	Dwyer et al. 1981a
Brook Trout	$\Delta L = -0.068 + 0.0578\ T$	7–16	0.882	Dwyer et al. 1981a
Rainbow Trout	$\Delta L = -0.040 + 0.0505\ T$	4–19	0.886	Dwyer et al. 1981b
Rainbow Trout	$\Delta L = 0.043 + 0.0450\ T$	7–19	0.801	Dwyer et al. 1981b
Rainbow Trout	$\Delta L = -0.167 + 0.066\ T$	7–16	0.971	Dwyer et al. 1981b
Lake Trout	$\Delta L = 0.176 + 0.0426\ T$	4–16	0.858	Dwyer et al. 1981c
Lake Trout	$\Delta L = 0.0622 + 0.0588\ T$	4–13	0.979	Dwyer et al. 1981c
Steelhead Trout	$\Delta L = 0.0329 + 0.0294\ T$	4–19	0.856	Dwyer et al. 1982
Steelhead Trout	$\Delta L = 0.0148 + 0.0343\ T$	7–16	0.937	Dwyer et al. 1982
Atlantic Salmon	$\Delta L = 0.0043 + 0.0306\ T$	4–19	0.926	Dwyer and Piper 1987
Atlantic Salmon	$\Delta L = -0.0429 + 0.0371\ T$	7–16	0.999	Dwyer and Piper 1987
Channel Catfish	$\Delta L = 0.612 + 0.0298\ T$	24–30	0.825	Andrews et al. 1972
Channel Catfish	$\Delta L = 0.195 + 0.0463\ T$	24–28	0.991	Andrews et al. 1972
Tiger Muskellunge				
3–4 cm	$\Delta L = -0.0548 + 0.0912\ T$	14–24	0.985	Meade et al. 1983
12–13 cm	$\Delta L = -0.0394 + 0.0471\ T$	18–24	0.864	Meade et al. 1983
Blue Tilapia	$\Delta L = 0.853 + 0.048\ T$	20–30	>0.99	Soderberg 1990

Note: T = temperature in °C and ΔL = daily growth in mm/day; r^2 is the coefficient of determination.

From Soderberg, R. W., *Progressive Fish-Culturist,* 54, 257, 1992. With permission.

Haskell (1959) determined the value of K to be 0.0004 for brook, brown, and rainbow trouts when W is given in pounds and L in inches. Representative average K values of some other species of cultured fish are listed in Table 2.11.

An example of fish weight projection follows. Suppose a hatchery manager wishes to estimate the monthly gain of 20,000 6-in. channel catfish being reared at 28°C. The Centigrade MTU required per centimeter of catfish length gain are 6.3 (Table 2.6). Thus, the expected monthly growth is 28/6.3 = 4.4 cm, or 1.75 in., and the fish should be 7.75 in. long at the end of the month. The weight of a 6-in. catfish is calculated to be $W = (0.0002877)(6)^3 = 0.062$ lb, and the weight of a 7.75-in. catfish is $W = (0.0002877)(7.75)^3 = 0.134$ lb. The expected gain per fish is $0.134 - 0.062 = 0.072$ lb, and the expected gain of 20,000 fish is 1440 lb.

**Table 2.11. Average K Values for Some Species
of Cultured Fish**

Species	$K \times 10^4$	Ref.
Brook, Brown, and Rainbow Trout	4.055	Haskell 1959
Muskellunge	1.600	Piper et al. 1982
Northern Pike	1.811	Piper et al. 1982
Lake Trout	2.723	Piper et al. 1982
Chinook Salmon	2.959	Piper et al. 1982
Walleye	3.000	Piper et al. 1982
Channel Catfish	2.877	Piper et al. 1882
Cutthroat Trout	3.559	Piper et al. 1982
Coho Salmon	3.737	Piper et al. 1982
Steelhead	3.405	Piper et al. 1982
Largemouth Bass	4.606	Piper et al. 1982
Blue Tilapia	8.430	Soderberg 1990

Note: $K = W/L^3$, where W = weight in pounds and L = length in inches.

Regression Model Procedure

Regression equations provide a less confusing method for the calculation of ΔL than temperature unit procedures. They should also be more accurate than temperature unit models that do not correct for the zero-growth intercept. The first step in the use of these equations is to select the one with the highest coefficient of determination whose temperature range includes the temperature for which ΔL is required. For example, if the growth of rainbow trout at 10°C were required, the equation $\Delta L = -0.167 = 0.666$ ($r^2 = 0.971$) should be used because the other two equations for this species were determined over greater temperature ranges and thus have lower coefficients of determination. Once the ΔL value is calculated it is used in the projection of fish growth as previously demonstrated.

When fish are inventoried, the projected length is compared to the actual length and the ΔL value is adjusted to account for any hatchery-specific deviations from the growth model.

FEEDING INTENSIVELY CULTURED FISH

The nutritional requirements of fish are not completely known, but adequate commercial diets are available for trout, salmon, and catfish. These species are carnivorous and require high-protein diets because they digest carbohydrates rather poorly compared to other livestock. Available Calories per gram of protein, fat, and carbohydrates for trout are approximately 3.9, 8.0, and 1.6, respectively (Piper et al. 1982). Approximately 3850 are required per kilogram of trout weight gain (Piper et al. 1982). Thus the food conversion, or the amount of food required per unit of fish weight gain, can easily be estimated if the proximate analysis of the diet is known. For example, suppose a trout diet contains 45% protein, 8% fat, and 10% carbohydrate. The available Calories in a kilogram of this diet are

Protein	450 g/kg × 3.9 C/g =	1750 C/kg
Fat	80 g/kg × 8.0 C/g =	640 C/kg
Carbohydrate	100 g/kg × 1.6 C/g =	160 C/kg
Total		2550 C/kg

The anticipated food conversion then is

$$\frac{3850 \ \text{C/kg fish}}{2550 \ \text{C/kg food}} = \frac{1.51 \ \text{kg food}}{\text{kg gain}}$$

Channel catfish, and probably other warm-water fish, digest components of their diet differently than trout. Available Calories per gram of protein, fat, and carbohydrate are 4.5, 8.5, and 2.9, respectively. Approximately 2156 are required per kilogram of catfish gain (Piper et al. 1982).

The food conversion is most commonly determined for a particular diet and fish strain from past station records. Calculation of the estimated food conversion is useful for evaluating potential diets, beginning a feeding schedule for new hatcheries, or investigating poor feeding efficiency.

Haskell (1959) introduced a feeding equation

$$F = \frac{3 \times C \times \Delta L \times 100}{L} \qquad\qquad 2.2$$

where F = percent body weight to feed per day, C = food conversion, DL = daily increase in length, and L = fish length on day fed. Notice that the feed rate equation calculates the amount of food required to result in a particular increment of growth. Normally, the DL value will be the maximum potential growth of a particular fish species at a certain growth temperature. If production schedules call for slower growth rates to result in a particular size of fish on a given date, a smaller DL value may be substituted.

Buterbaugh and Willoughby (1967) noted that when the diet, fish strain, and water temperatures were unchanged, the factors in the numerator of the feed rate equation remained constant. They introduced a hatchery constant, HC, which combines these factors in a single term. Thus,

$$HC = 3 \times C \times \Delta L \times 100 \qquad\qquad 2.3$$

and

$$F = \frac{HC}{L} \qquad\qquad 2.4$$

HATCHERY RECORDS

Daily food rations for each lot of fish are calculated and growth is projected each day by adding ΔL to L. Usually a sample count is taken approximately every three months when the fish are graded. At this time the food conversion can be calculated and the projected fish size can be corrected with the actual value obtained in the sample count.

SAMPLE PROBLEMS

1. You know from past station records that 1.7 pounds of feed are required for each pound of fish gain. Your facility holds 20,000 9-in. rainbow trout with a K factor of 0.0004. You are assigned to write up the feed order for July, August, and September. The hatchery manager tells you that the expected water temperatures for these months are 12, 13, and 11°C, respectively. How much feed should you order?

2. A hatchery manager is requested to supply 9.5-in. fish on April 1 for the opening day of trout season. The water temperature at this hatchery is a constant 12°C. When should he have eggs delivered assuming that they hatch as soon as they arrive and the fry are 0.8 in. long at that time? Assume a 2-week swim up time.

3. Calculate the feed cost of producing a 9.5-in. brook trout given the following information.

   Length at first feeding is 0.6 in.

   K = 0.0004

   Food conversion is 1.3

   Food cost is $430/ton.

4. Tiger muskies, the cross between a muskellunge and northern pike, are now grown routinely in hatcheries on dry feed. Newly hatched fry 0.4 in. long can reach a length of 9 in. in the three-month summer season. What is the hatchery constant for a hatchery experiencing this growth rate and a feed conversion of 1.7?

5. For the above example, how much feed is required to raise 100,000 9-in. tiger muskies? K = 0.00016.

6. You are assigned to feed fish at a trout hatchery. Given the following information, how much feed do you require on August 15?

   Lot No. 1: Sample count August 1, 946 fish weighed 10.52 lb. Total fish in lot, 22,501.

   Lot No 2: Sample count August 1, 501 fish weighed 43.29 lb. Total fish in lot, 19,055.

   Lot No. 3: Sample count August 1, 200 fish weighed 195 lb. Total fish in lot, 51,250.

   Hatchery constant = 8.55. Last year this hatchery produced 150,000 pounds of fish on 198,000 pounds of feed. K − 0.0004.

7. For the following lot of fish calculate:
   a) the hatchery constant,
   b) the feed ration for August 1,
   c) the feed ration for August 30.
   Estimated feed conversion = 1.5; water temperature = 53°F; sample count on
   July 15: 936 fish weighed 302 lb; K = 0.0004; lot contains 215,000 fish.
8. Calculate the feed requirements for July for the following lot of catfish being
   grown in a geothermal spring.
   Sample count June 1: 1,026 fish weighed 251 kg; water temperature is 83°F;
   K = 0.0002877; C = 1.7; lot contains 110,000 fish.
9. Choose the most economical diet given the following information.
   Feed A costs $370/ton and gives a feed conversion of 1.85
   Feed B costs $430/ton and gives a feed conversion of 1.50
   Feed C costs $485/ton and gives a feed conversion of 1.45
10. Choose the most economical diet from those listed below.

Diet	% Protein	% Fat	% Carbohydrate	Cost, $/kg
A	38	5	16	0.46
B	42	5	12	0.48
C	45	8	10	0.51
D	50	12	6	0.62

REFERENCES

Andrews, J. W., L. H. Knight, and T. Murai. 1972. Temperature requirements for high
    density rearing of channel catfish from fingerlings to market size. *Progressive
    Fish-Culturist* 34: 240-241.
Buterbaugh, G. L. and H. Willoughby. 1967. A feeding guide for brook, brown, and
    rainbow trout. *Progressive Fish-Culturist* 29: 210-215.
Dwyer, W. P. and R. G. Piper. 1987. Atlantic salmon growth efficiency as affected by
    temperature. *Progressive Fish-Culturist* 49: 57-59.
Dwyer, W. P., R. G. Piper, and C. E. Smith. 1981a. Brook Trout Growth Efficiency as
    Affected by Temperature. U.S. Fish and Wildlife Service Fish Culture Develop-
    ment Center (Bozeman, Montana) Information Leaflet 16.
Dwyer, W. P., R. G. Piper, and C. E. Smith. 1981b. Rainbow Trout Growth Efficiency
    as Affected by Temperature. U.S. Fish and Wildlife Service Fish Culture Devel-
    opment Center (Bozeman, Montana) Information Leaflet 18.
Dwyer, W. P., R. G. Piper, and C. E. Smith. 1981c. *Lake Trout* Salvelinus Namaycush
    Growth Efficiency as Affected by Temperature. U.S. Fish and Wildlife Service Fish
    Culture Development Station (Bozeman, Montana) Information Leaflet 22.
Dwyer, W. P., R. G. Piper, and C. E. Smith. 1982. *Steelhead Trout Growth Efficiency
    as Affected by Temperature*. U.S. Fish and Wildlife Service Fish Culture Devel-
    opment Center (Bozeman, Montana) Information Leaflet 27.
Haskell, D. C. 1959. Trout growth in hatcheries. *New York Fish and Game Journal* 6:
    205-237.

Meade, J. W., W. F. Krise, and T. Ort. 1983. Effect of temperature on production of tiger muskellunge in intensive culture. *Aquaculture* 32: 157-164.

Piper, R. G., J. B. McElwain, L. E. Orme, J. P. McCraren, L. G. Fowler, and J. R. Leonard. 1982. *Fish Hatchery Management*. U.S. Fish and Wildlife Service, Washington, D.C.

Soderberg, R. W. 1990. Temperature effects on the growth of blue tilapias in intensive culture. *Progressive Fish-Culturist* 52: 155-157.

Soderberg, R. W. 1992. Linear fish growth models for intensive aquaculture. *Progressive Fish-Culturist* 54: 255-258.

CHAPTER 3

Water Sources for Flowing Water Fish Culture

THE HYDROLOGIC CYCLE

All the rivers run into the sea, yet the sea is not full; unto the place from whence the rivers come, thither they return again — Ecclesiastes 1:7.

Mankind has long been aware of the hydrologic cycle — the processes of evaporation, transportation, condensation, precipitation, percolation, and run-off — by which water is recycled through the hydrosphere.

Evaporation occurs when radiant heat is absorbed and converted to kinetic energy in water molecules. The resulting molecular velocity causes some molecules to leave the water surface and enter the atmosphere. The energy lost in this process is the latent heat of vaporization, and the water surface cools. In addition to solar radiation, the rate of evaporation is affected by humidity, which determines the vapor pressure gradient, water turbidity, which may affect the amount of heat absorbed at the water surface, and most importantly, by wind, which carries away the saturated layer of air at the water surface, maintaining the vapor pressure gradient. Plants increase the surface area over which evaporation occurs. The evapotranspiration rate from aquatic plants may be twice as high as evaporation from the free water surface.

Condensation is caused by temperature changes in air which affects its ability to hold moisture. The dew point is the temperature at which air becomes saturated with water. As air rises, it expands and cools according to the fundamental gas laws (Chapter 5). When the dew point is reached, condensation occurs and clouds form. Thus clouds are made up of small droplets of liquid water too small to fall as rain. Precipitation occurs when ice crystals, dust, or sea salts act as hydroscopic nuclei upon which water droplets coalesce into raindrops. Air rises, causing precipitation, by convective warming, by moving wedges of cold air that cause frontal storms, or by orographic processes in which air rises over geographic features.

Some rainfall is intercepted by plants and other obstacles and may evaporate before reaching the soil surface. The remainder enters the soil by percolation or runs off the soil surface into streams. The percolation rate is affected by gravity, capillarity, soil moisture content, and the amount of air in the soil pores. The layer of soil that retains soil moisture is called the zone of rock fracture. When this zone is saturated with water, infiltration continues to the zone of rock storage where it is referred to as groundwater.

Stream flow consists of runoff, channel precipitation, and base flow, minus water losses to consumption, evaporation, and seepage. Base flow is the contribution to stream discharge from groundwater. The amount of precipitation that runs off in a particular watershed is referred to as the hydrologic response, and is expressed as a percentage. The hydrologic response is calculated from a hydrograph, which measures direct runoff above base flow following a precipitation event, and from rain gauges in the watershed, which measure total precipitation. Hydrologic response values may range from 1 to 75% and are related to soil permeability, bedrock porosity, slope, and vegetation. Mountainous areas with shallow soils and impermeable bedrock have high hydrologic responses because a relatively large fraction of the water that falls as precipitation enters streams. Low hydrologic responses occur in areas where most of the precipitation enters the groundwater due to low gradient, permeable soils, and porous bedrock such as limestone or shale.

Over 97% of the water on earth is in the oceans, and over 77% of the remaining freshwater is locked in ice caps and glaciers. Some intensive aquacultures involve pumping of seawater to rearing units on land, but the water supplies usually used for flowing water fish culture are relatively shallow groundwaters and surface waters, usually from streams. These sources comprise a relatively small fraction of the water circulating through the hydrologic cycle, and in the case of streams, a very transient one.

USE OF SURFACE WATERS FOR FISH CULTURE

In areas of high hydrologic response, streams may be diverted through rearing units for flowing water fish culture. Stream discharge is composed of runoff and base flow, but in areas where a significant base flow occurs in streams, groundwater would probably be chosen as the aquaculture water supply. Thus, use of surface waters is most practical in locations that receive regular and heavy rainfall and have high hydrologic responses.

Temperatures of streams with low base flows vary seasonally with the air temperature, and the degree of variation depends mostly upon the magnitude of the base flow component and summer shading by riparian vegetation. Thus, a moderate climate and wooded watersheds are required for cold-water fish culture using stream water supplies, and a tropical climate is necessary for similar warm-water fish culture. Streams are used as water supplies for trout culture in the southern Appalachian region of the U.S. Suitable locations are higher than about 3,000 feet above sea level, because below this elevation, summer water temperatures may get too warm for trout survival and the winter period of reduced fish growth is excessive. Even at higher elevations, the growing season for trout is limited to about eight months due to high summer and low winter temperatures. A few experimental catfish farms in the southern U.S. have used reservoir water in flowing water designs, which is pumped back to the reservoir after passage through the rearing units. This technology should be economically comparable to traditional static-water catfish culture. Stream waters have not been used for intensive warm-water fish culture in temperate climates because they achieve temperatures suitable for satisfactory fish growth only for brief summer periods. Flowing water fish culture technology is not likely to be appropriate for tropical regions where surface water temperatures remain in the acceptable range for warm-water fish culture because of the costly complete diets required.

Environmental regulations in some states prohibit stream diversion, thereby precluding their use for fish culture.

USE OF GROUNDWATER FOR FISH CULTURE

Groundwater is that water contained in geologic material. The zone of saturation in water-bearing strata is called an aquifer. Aquifers are characterized by their porosity, which is related to the void fraction of the stratum, and their permeability, which describes the degree of water movement through the aquifer and is related to the pore size. Water-bearing strata may be well-sorted, such as sand, or a poorly sorted aggregate of different sized particles. In some aquifers mineral deposits may partially fill the voids. In rock aquifers water may be held in channels where the rock has dissolved or in voids created by rock fracture. The amount of water in saturated strata may range from 0.01 to 0.35% by volume.

Groundwater is usually chosen for flowing water fish culture because it has a nearly constant temperature throughout the year. This is because the temperature of the aquifer is maintained at approximately the mean temperature of the water entering it. It follows that the temperature of groundwater can be estimated from the annual mean air temperature of a region. This generalization holds true as long as groundwaters are not influenced by geothermal activity. In temperate climates groundwaters are much warmer than surface waters in the winter and much cooler than surface waters in the summer. For this reason, groundwater is our most useable source of solar energy. Average groundwater temperatures in the U.S. are shown in Figure 3.1. Note that the isotherms follow latitudinal as well as elevational determinants of mean air temperature. Thus, equatorial areas at high elevations, such as those found in South America and Africa, often have suitable groundwater as well as surface water temperatures for cold-water fish culture.

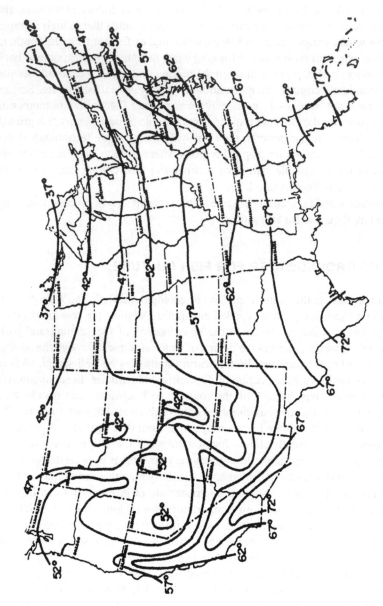

Figure 3.1. Groundwater isotherms for the U.S. (From Anon., *Ground Water and Wells*, Edward E. Johnson, Inc., St. Paul, MN. With permission.)

Aquifers are recharged by precipitation that percolates through the soil to the water-bearing stratum, or in some cases by streams that flow into permeable formations. Groundwater may be accessed at springs, where it erupts through the surface or by wells driven into the aquifer. If a well is driven into an unconfined aquifer, i.e., one where the water is not under pressure, water will fill the pipe to the upper surface of the aquifer, or water table, and must be pumped. In a confined aquifer, the water is under pressure caused by the head differential from the recharge area to the well. If this aquifer is accessed by a well, water will rise in the pipe to the piezometric surface, which is a level determined by the amount of head differential and the frictional head losses that occur through the aquifer. Such wells are called artesian and may be desirable for aquaculture water supplies because the pumping depth may be less than that from wells in unconfined aquifers. If the piezometric surface is above ground level, water will flow from an artesian well without pumping. Discharges from flowing wells are sometimes increased by pumping.

When a stream bed intersects an aquifer a considerable portion of its discharge may be base flow. If the stream bottom lies above the water table, water seeps out of the stream toward the aquifer. The resulting elevated mound of groundwater is called bank storage and may extend for several miles from the beds of large rivers.

When a well is pumped, the area around it becomes de-watered and the water level drops. When equilibrium between discharge of the well and recharge of the pumped area from the aquifer is reached there will be a cone-shaped de-watered area around the well. This area is called the cone of depression and its bottom is the actual depth from which water must be pumped. The dimensions of the cone of depression are determined by the permeability of the aquifer and the recharge rate. Wells drilled into highly permeable strata will have wide, shallow cones of depression while those in low permeability strata will have narrow, deep cones of depression resulting in a considerable drawdown during pumping and consequent increase in pumping depth. Because pumping cost is related to the depth from which water is pumped, the dimensions of the cone of depression are an important consideration in the economic evaluation of a well for aquaculture. A test well must be drilled and operated in order to determine the actual pumping depth from a particular aquifer. In areas where wells are closely spaced, cones of depression may overlap, causing a localized lowering of the water table.

The pump size required for a particular job is calculated from the conversion factor of 33,000 ft·lb/min of work per horsepower (HP). Suppose we wish to pump 1000 gpm from 100 ft. The horsepower requirement is

$$\frac{1000 \text{ gpm} \times 8.34 \text{ lb/gal} \times 100 \text{ ft}}{33,000 \text{ ft} \cdot \text{lb/min/HP}} = 25 \text{ HP}$$

The efficiency of the pump engine describes the relationship between water HP, which we have just calculated and brake HP, which the pump engine must develop. If the motor is 75% efficient, typical of electrically powered units, 25/0.75 = 33 HP, which is the amount of power that the pump requires. Pumping cost is calculated by converting HP to kW. For the present example, using electrical power purchased at $0.07/kWh, the annual pumping cost is

$$33 \text{ HP} \times \frac{0.746 \text{ kW}}{\text{HP}} \times \frac{8640 \text{ hr}}{\text{year}} \times \frac{\$0.07}{\text{kWh}} = \$14,889/\text{year}$$

The pumping cost for a diesel pump is determined as follows, assuming that the engine is 50% efficient and fuel contains 110,000 BTU/gal and is purchased for $1.00/gal:

$$\frac{25 \text{ HP}}{0.50} \times \frac{2545 \text{ BTU}}{\text{HPh}} \times \frac{\text{gal}}{110,000 \text{ BTU}} \times \frac{8640 \text{ hr}}{\text{year}} \times \frac{\$1.00}{\text{gal}} = \$9,995/\text{year}$$

WATER FLOW IN CHANNELS AND PIPES

Water Flow in Open Channels

Water for flowing water fish culture must often be channeled and piped from its source to a suitable location for the rearing units. Flow rates through open channels are estimated by first calculating the expected velocity and then calculating the discharge from

$$Q = VA \qquad\qquad 3.1$$

where Q = discharge in units cubed per unit time, V = velocity in linear units over time, and A = cross-sectional area of flow in units squared.

Manning's equation is generally used to estimate water velocities in open channels:

$$V = \frac{1}{n} R^{2/3} S^{1/3} \qquad\qquad 3.2$$

where V = velocity in m/sec, n = roughness coefficient of the channel walls, R = hydraulic radius in m, and S = channel slope in units of fall per unit of length. The hydraulic radius is a measure of the geometric efficiency of the cross section and is equal to the cross-sectional area divided by the wetted perimeter. Thus, for a rectangular channel, R = hb/2h + b, where h is the height and b is the length of the base. For a trapezoidal channel, R = hb/2h tan (ϑ/2) where ϑ = the angle of the channel's side. Values for n have been determined empirically and are found in Table 3.1.

Consider the flow through a winding earthen channel whose cross section most resembles a half-circle with a radius of 0.2 m. Thus, the cross-sectional area is $\pi(0.2)^2/2 = 0.063$ m^2, and the wetted perimeter $= \pi(0.4)/2 = 0.63$ m. The head of the channel is 200 m from its end, over which the elevation difference is 0.4 m. Let us estimate the amount of water this channel can deliver to its downstream end. The value for n is estimated at 0.026 from Table 3.1. R = 0.063/0.63 = 0.1. From Equation 3.2

$$V = \frac{1}{0.026}(0.1)^{2/3}\left(\frac{0.4}{200}\right)^{1/3} = 1.0 \text{ m/sec}$$

From Equation 3.1, Q = 1.0 × 0.063 = 0.063 m^3/sec.

For soil- and vegetation-lined channels, erosion and sedimentation must be considered. Excessive velocities will erode the channel and low velocities of water containing suspended material will cause it to be filled in. Maximum velocities in bare-earth channels range from 0.5 to 1.5 m/sec, depending upon the soil type. Channels lined with vegetation can carry greater flow than can channels without vegetation.

Water Flow in Pipes

The conveyance method is a useful and simple procedure for determining pipe sizes and discharges for piped water systems. The conveyance of a particular pipe is calculated as,

$$K = \left[6.304\left(2 \log \frac{D}{e} + 1.14\right)D^{2.5}\right] \qquad 3.3$$

where K = conveyance, D = pipe diameter in ft, and e is a frictional coefficient related to the roughness of the pipe material. Values for e of some pipe surfaces are provided in Table 3.2. Discharge is calculated from conveyance by

$$Q = K\sqrt{\frac{h_L}{L}} \qquad 3.4$$

where Q = discharge in cubic feet per second (cfs), h_L = head loss in ft, and L = the length of the pipe in ft.

Head loss is the elevational difference between the inlet and outlet of the pipe. Energy is also lost by water flowing through pipeline systems due to passage through fittings and valves. These losses are called local losses and are

Table 3.1. Roughness Coefficient, n, for Use in
Manning's Formula

Type of channel	n Value range
Earth channels	
Earth, straight and uniform	0.017–0.025
Earth, winding and sluggish	0.022–0.030
Earth bottom, rubble sides	0.028–0.033
Dredged earth channels	0.025–0.035
Rock cuts, smooth and uniform	0.025–0.035
Rock cuts, jagged and irregular	0.035–0.050
Vegetated channels	
Dense, uniform, over 10 in. long	
Bermuda grass	0.040–0.200
Kudzu	0.070–0.230
Lespedeza	0.047–0.095
Dense, uniform, less than 2.5 in. long	
Bermuda grass	0.034–0.110
Kudzu	0.045–0.160
Lespedeza	0.023–0.050
Natural streams	
Straight banks, uniform depth	0.025–0.040
Winding, some pools, few stones or weeds	0.033–0.055
Sluggish with deep pools, stones or weeds	0.045–0.060
Very weedy streams	0.075–0.150
Lined channels	
Concrete-lined	0.012–0.018
Smooth metal	0.011–0.015
Corrugated metal	0.021–0.026
Wood flumes	0.010–0.015

From Wheaton, F. W., *Aquaculture Engineering*, John Wiley &
Sons, New York, 1977. With permission.

Table 3.2. Roughness Coefficient, e, for Use
in the Conveyance Equation

Pipe surface	Roughness
Glass, plastic, etc	Smooth*
Wrought iron, steel	1.5×10^{-4}
Galvanized iron	5.0×10^{-4}
Cast iron	8.5×10^{-4}
Concrete	$1.1 \times 10^{-3} - 1 \times 10^{-2}$

* For smooth pipes, e/D is approximately 1×10^{-6},
where D = the pipe diameter in ft.
Modified from Simon, A. L., *Practical Hydraulics*,
John Wiley & Sons, New York, 1979. With permission.

most conveniently expressed as equivalent pipe lengths. These values, in pipe
diameters, are provided in Table 3.3.

An example using these procedures follows: Calculate the discharge from
a 3/4-in. steel pipe that is 100 ft long and contains two 90°-elbows and an open
gate valve. The elevational difference (h_L) between the inlet and outlet ends of
the pipe is 3 ft.

Table 3.3. **Equivalent Pipe Length of Local Head Losses in Fittings and Valves, Expressed in Number of Pipe Diameters**

Fitting or valve	Equivalent length (diameters)
Tee (run)	20
Tee (branch)	60
90°-elbow, short radius	32
90°-elbow, long radius	20
45°-elbow	15
Gate valve (open)	17
Butterfly valve (open)	40

From Hammer, M. J., *Water and Waste-Water Technology*, John Wiley & Sons, New York, 1977. With permission.

From Equation 3.3, the conveyance of the pipe is

$$K = \left[6.304 \left(2 \log \frac{0.0625}{1.5 \times 10^{-4}} + 1.14 \right) 0.0625^{2.5} \right] = 0.039 \text{ cfs}$$

The local losses in pipe diameters are 32 for each of the elbows and 17 for the valve. Thus, the total pipe length equivalency of these devices is 0.0625 (32 + 32 + 17) = 5.1 ft. From Equation 3.4,

$$Q = 0.039 \sqrt{\frac{3}{105.1}} = 0.007 \text{ cfs} = 2.96 \text{ gpm}$$

Most hydraulic engineers would consider the procedure outlined above to be an oversimplification of piped water system design. For the text for a course in hydraulics this is certainly true. In actuality, the complexities of the interactions between flow, velocity, pressure, and pipe and fitting frictional losses are so great that pipeline design becomes a trial and error process of calculative iteration. The procedure described here is intended to give the student and practitioner of aquaculture a first-order approximation of pipe sizes needed to carry required flows.

WATER PUMPS

A centrifugal pump, which is appropriate for most aquaculture applications, is a cavity with an inlet and outlet for water flow in which an impeller spins to force the water in the desired direction. Pumps designed to move small volumes of water to great heights have large impellers and small cavities, while pumps designed to move large water volumes over low head pressures have small impellers turning in large cavities. Pump performance is quantified by the term specific speed computed from

$$n_s = \frac{\text{rpm} \sqrt{\text{gpm}}}{H^{0.75}}$$ 3.5

where n_s = specific speed in revolutions per minute (rpm), rpm is the rotational speed of the shaft, gpm is the discharge in gal/min, and H is the total head in ft. Pumps with low specific speeds (500 to 2000 rpm) are designed for small discharges and high pressures while pumps designed to deliver large discharges over low heads have high n_s values (5000 to 15000 rpm).

Suppose we wish to lift 1500 gpm of water over a 4-ft levy to supply water from a stone quarry to a trout production facility. We have an engine that runs at 1000 rpm. Ignoring head losses in the pipe and fittings,

$$n_s = \frac{1000 \sqrt{1500}}{4^{0.75}} = 13693 \text{ rpm}$$

We require a pump with a specific speed of 13,693 rpm, which is within the range of low-head, high-volume pumps. Diagrams showing the particular pump design required for a given performance characteristic are available from pump manufacturers.

Pump sizing procedures must account for head losses due to pipe friction and fittings as previously described.

MEASUREMENT OF WATER FLOW RATE

Water flow in pipes can be measured with a commercially available in-line device. For many aquaculture applications, a stopwatch and container of known volume will provide a sufficiently accurate measure of pipe discharge.

Measurement of flow through an open channel is usually accomplished by forcing the flow through a weir that is rectangular, trapezoidal, or 90°-triangular in shape. Smaller flows are best measured with a triangular weir. Care must be taken to assure that the entire water flow passes through the weir, rather than around the sides or bottom. For a rectangular weir,

$$Q = 1.84 \text{ Lw } (h)^{0.67}$$ 3.6

where Q = discharge in m³/sec, Lw = the width of the weir and h = the height that the water rises in the weir at equilibrium. For best accuracy the difference between the stream width and the width of the weir should exceed by a factor of four the water height in the weir. For a trapezoidal weir with a 4:1 side slope,

$$Q = 1.37 \, Lw \, (h)^{0.67} \qquad\qquad 3.7$$

and for a triangular weir,

$$Q = 1.37 \, Lw \, (h)^{2.5} \qquad\qquad 3.8$$

SAMPLE PROBLEMS

1. A reservoir is constructed that traps all the runoff from a 1000-acre watershed with a hydrologic response of 10% and an average annual rainfall of 40 in./year. A flowing water aquaculture facility is constructed below the dam using the reservoir as a water supply. Calculate the constant water flow available in gal/min.

2. A rainfall event results in a average 0.3 in. precipitation over a 50,000-acre watershed. The total discharge, above base flow, from the hydrograph is 5,000,000 ft³. Calculate the hydrologic response of this watershed.

3. Estimate the maximum amount of water that can be safely pumped from an aquifer whose recharge basin is 100 square miles and has an annual rainfall of 20 in./year and a hydrologic response of 35%. Your answer should be in gpm.

4. Calculate the annual cost to pump 1000 gpm from a well using an electric pump. The depth to the water table is 100 ft and the drawdown is 20 ft. Use $0.065/kWh for the cost of electricity and 80% for the pump efficiency.

5. A 100-HP diesel-powered pump has a discharge of 1200 gpm from an unconfined aquifer whose water table is at a depth of 120 ft. What is the drawdown?

6. Design a canal to carry 16 cfs of water from a spring to an aquaculture facility. A soil engineer informs you that the channel should be trapezoidal in cross section with a 2:1 side slope and flow velocities must not exceed 0.5 m/sec.

7. Use the conveyance method to calculate the correct size of plastic pipe for a pipeline containing two 45°-elbows, a run tee, a 90°-elbow, and an open gate valve. The length of the line is 200 ft, the available head is 1.5 ft, and the required discharge is 50 gpm.

8. Calculate the discharge, in gpm, through a rectagular concrete sluice that is 2 ft wide, 1 ft deep, and has a slope of 1:1000.

9. A catfish hatchery requires a pump to supply 50 gpm of groundwater from a well whose pumping depth is 175 ft. The pump motor rotates at 1800 rpm. Calculate the specific speed and horsepower of the pump that should be selected.

10. A stream is channeled through a trapezoidal weir with a top width of 0.5 m in order to measure its volume of flow. What is the stream discharge if the water height in the weir is 20 cm?

REFERENCES

Anon. 1975. *Ground Water and Wells*. Johnson Division, UOP Inc., Edward E. Johnson, Inc. St. Paul, MN.

Hammer, M. J. 1977. *Water and Waste-Water Technology*. John Wiley & Sons. New York, NY.

Simon, A. L. 1979. *Practical Hydraulics*. John Wiley & Sons. New York, NY.

Todd, D. K. 1964. *Ground Water Hydrology*. John Wiley & Sons. New York, NY.

Wheaton, F. W. 1977. *Aquaculture Engineering*. John Wiley & Sons. New York, NY.

Yoo, K. H. and C. E. Boyd. 1993. *Hydrology and Water Supply for Pond Aquaculture*. Chapman Hall. New York, NY.

CHAPTER 4

Fish Culture Rearing Units

Flowing water fish culture is conducted in one of two general types of rearing units that differ in their hydraulic characteristics. Linear units, usually called raceways (Figures 4.1 to 4.3), exhibit a hydraulic pattern that approximates plug flow in which all elements of the water move with the same horizontal velocity. Circulating units, such as round tanks (Figure 4.4), have nonuniform velocities and new water is added to a mass of circulating used water. Thus, linear units containing fish exhibit a water-quality gradient from the inlet to the outlet while circulating units have relatively uniform water quality at all positions.

LINEAR UNITS

Haskell et al. (1960) reported unsatisfactory fish production in raceways that had throated inlets and outlets. This was due to dead spots in the pond resulting from poor water circulation. By injecting dye into the pond influent they showed that the water flowed along one side of the raceway, leaving a dead area on the opposite side. When water entered and exited the raceway over its entire width, plug-flow was achieved and fish production improved (Haskell et al. 1960).

Salmonid production raceways (Figure 4.5) are usually constructed of poured, reinforced concrete and generally have a length:width ratio of 10:1. Buss and Miller (1971) recommended that raceway depths not exceed 2 ft. Westers and Pratt (1977) recommended that raceways have a water exchange rate of 4 exchanges/hr and a linear velocity of 0.033 m/sec. An exchange rate of 4 exchanges/hr was deemed to be the best compromise between required pond size, velocity, and waste removal. A velocity greater than 0.033 m/sec will not allow solid waste to settle out at the rear, and a lower velocity will cause solids to settle within the pond, among the fish, making cleaning more difficult (Figure 4.6) (Westers and Pratt 1977).

31

Figure 4.1. Linear fish-rearing units in Spearfish, South Dakota, used to produce trout for stocking.

Figure 4.2. This commercial hatchery in southern Idaho uses large raceways for the production of rainbow trout.

Figure 4.3. Pumped groundwater supplies these raceways used to produce tilapia in Kenya.

Figure 4.4. Circular rearing units constructed from poured, reinforced concrete are used to rear rainbow trout at this farm in Montana.

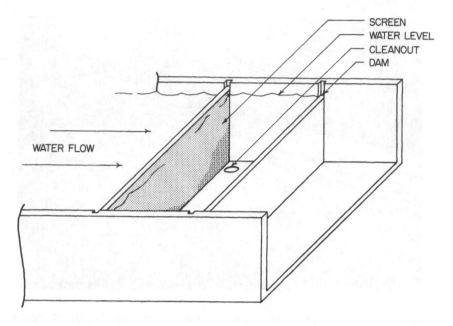

Figure 4.5. Detail of discharge of a typical salmonid production raceway.

Figure 4.6. A raceway velocity of 0.033 m/sec allows solid waste to settle in the fish-free area between the screen and the dam boards from where it can be removed through a clean-out drain.

Raceways are generally built in series of multiple sections with a clean-out plug at the end of each section. The clean-outs are connected to a drainage line that leads to a settling pond. The number of sections that may be successfully placed in a series depends upon the reaeration capability between them (Chapter 8) and the accumulation of dissolved metabolites (Chapter 9).

The following example illustrates the process of raceway design. Suppose a raceway is to be designed to receive 20 L/sec of water. The length:width ratio should be approximately 10:1, the desired depth is 0.2 m, the flow velocity must be 0.033 m/sec, and the exchange rate should be 4 exchanges/hr. The total volume of the raceway is

$$\frac{20 \text{ L}}{\text{sec}} \times \frac{3600 \text{ sec}}{\text{hr}} \times \frac{\text{hr}}{4 \text{ exchanges}} \times \frac{\text{m}^3}{1000 \text{ L}} = \frac{18 \text{ m}^3}{\text{exchange}}$$

Since the velocity and depth are set, the width must be calculated. Discharge, velocity, and the cross-sectional area of the raceway, width (W) × depth (D), are related according to Q = VA (Equation 3.1). The flow is set at 20 L (0.020 m³)/sec and the depth is set at 0.2 m. Thus,

$$0.020 \text{ m}^3 = W \times 0.2 \times 0.033$$

and

$$W = \frac{0.020}{0.2 \times 0.03} = 3.0 \text{ m}$$

The total pond volume must be 18 m³, the depth is 0.2 m, and the width is 3.0 m. Thus, the length (L) is

$$L = \frac{18}{3.0 \times 0.2} = 30 \text{ m}$$

Notice that the requirement for a length:width ratio of 10:1 is met. An alternative raceway design method is to set the dimensions of the raceway at 30:3:1 (L:W:D). If a flow velocity of 0.033 m/sec for a 0.020 m³/sec flow is desired, set D at x so that W = 3x. Then,

$$0.020 = 3x \times x \times 0.033$$

$$x^2 = \frac{0.020}{3 \times 0.033}$$

$$x = 0.45 \text{ m} = D$$

and

$$W = 3x = 1.35 \text{ m}$$

A 10:1 length:width ratio is achieved by making the raceway 13.5 m in length. The total pond volume, then, is 8 m³ and the exchange rate is 9 exchanges/hr. This raceway also has a satisfactory design. The preceding examples show that either depth, exchange rate, or dimensional ratios may be manipulated when designing a raceway with particular velocity and dimensional requirements.

Buss et al. (1970) and Moody and McCleskey (1978) described the use of vertical raceways, or silos, in which water is injected at the bottom of a large vertical tube, and allowed to exit at the top. Buss et al. (1970) first used 55-gal steel drums with a water capacity of 43 gal. When the drums were supplied with 3 to 6 gpm of water, waste accumulated at the bottom, but at 12 gpm, the units were self-cleaning. Self-cleaning drums were not advised if placed in series because of the accumulation of suspended solid waste in downstream units. Buss et al. (1970) also reported successful trout production in fiberglass silos that were 16.5 ft tall, 7.5 ft in diameter, and supplied with 450 gpm of water. These units were self-cleaning so, presumably, a lower flow rate would be recommended if several were placed in series. The water exchange rate in the silo was 5 exchanges/hr, which allowed fish densities to exceed those typical of linear raceways.

Moody and McCleskey (1978) described trout production in a serial silo system. The units were 11 ft high, 7.25 ft in diameter, and supplied with 300 gpm of water. Thus, the exchange rate was 5.3 exchanges/hr. There were five silos in the series with a 3-ft drop between them and a clarifier between the fourth and fifth units. Fish growth in the silos was better than in adjacent raceways and fish densities were three times greater than in the raceways because of the more rapid water-exchange rate. Water quality was better through five successive uses in the silos than when reused similarly through raceways in series. Consequently, fish production was better in the silo series than in the raceway series.

The major advantage of vertical raceways over horizontal units is that much less physical space is required for the former (Buss et al. 1970). Fish feeding is probably less time-consuming in silos than in raceways (Buss et al. 1970; Moody and McCleskey 1978). A disadvantage of vertical units is the difficulty in fish removal for sampling, grading, or harvest.

CIRCULATING UNITS

Circulating fish-rearing units are usually circular in shape with a center drain. Water is admitted, usually under pressure, at a single location along the edge. Circular units are usually 2 to 3 ft deep and 3 to 50 ft in diameter.

Burrows and Chenoweth (1955), and Larmoyeaux et al. (1973) studied the hydraulic characteristics of circular ponds. Both studies revealed a donut-shaped dead area in the center of the unit where water circulation was poor (Figure 4.7). Because of this dead area, mid-depth velocities were greatest at the edge, intermediate near the outlet, and lowest in the center of the unit (Burrows and Chenoweth 1955). The size of the dead area and degree of stagnation can be reduced by increasing the depth of the pond or increasing the pressure at which water is injected (Burrows and Chenoweth 1955; Larmoyeaux et al. 1973). Circular ponds are self-cleaning because the general decrease in velocity from the outside toward the center sweeps solid waste toward the drain. The donut-shaped dead zone is directly above an area with relatively high velocity so that debris settling there is swept away (Burrows and Chenoweth 1955).

Design criteria for circular units are not available, but generally the water exchange rate does not exceed 2 exchanges/hr. Thus, fish densities in circular ponds are lower than in most linear units.

Burrows and Chenoweth (1970) described a rectangular circulating pond with a center wall (Figure 4.8). Water is injected under pressure at two locations on opposite ends of the center wall. Drains are located at opposite ends of the center wall and are flush with the pond floor. The flow pattern around the pond is controlled by turning vanes at each corner. Water flows parallel to the pond walls and gradually reaches the center wall as the velocity decreases. The pond is self-cleaning and contains no stagnant zones (Burrows and Chenoweth 1970).

Circular ponds have flow velocities five to ten times greater than raceways (Burrows and Chenoweth 1955), which could be advantageous when rearing fish for increased stamina to improve survival after stocking. On the other hand, increased swimming speed increases oxygen demand, which lowers carrying capacity and increases food conversion rates, increasing the cost of production. Larmoyeaux et al. (1973) reported that fish loads in circular ponds could be higher than in raceways receiving the same amount of water because dissolved oxygen is added to the unit by the high-pressure inflow. However, injecting water under pressure is not an energy-efficient means of aeration (see Chapter 8). Burrows and Chenoweth (1955) doubted that circular ponds could hold more fish than raceways receiving equal amounts of water. Buss and Miller (1971) and Westers and Pratt (1977) reported that a water-quality gradient, as provided by linear rearing units, was preferable to uniform water quality throughout the unit and recommended raceways over circulating ponds for salmonid production.

SAMPLE PROBLEMS

1. Design a raceway to receive 600 gpm of water, have the dimensional ratios of 30:3:1 (L:W:D), and have four water exchanges per hour.

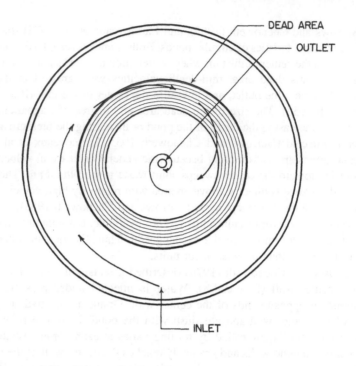

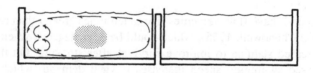

Figure 4.7. Hydraulic characteristics of a circular fish-rearing unit. [From Burrows and Chenoweth (1955) and Larmoyeaux et al. (1973).]

2. What is the linear velocity of water in the unit you designed above? Alter the raceway design so that the velocity is 0.1 ft/sec, and still provides 4 water exchanges per hour. Alter the raceway design so that the flow velocity is still 0.1 ft/sec, and the dimensional ratios are 30:3:1. Which of the three raceway designs is best for fish culture? Why?

3. Raceways used for catfish production in Idaho are 24 ft long, 8 ft wide, and 3 ft deep. There is one water exchange every 6 min. What is the flow rate and linear velocity? How could the velocity be changed to 0.1 ft/sec? What would be the new water exchange rate?

4. What is the water exchange rate and vertical velocity of a 55-gal drum with a water capacity of 43 gal, supplied with 12 gpm of water? The drum diameter is 22 in.

5. Five water exchanges per hour are required for a vertical raceway that is 3 m in height and 2 m in diameter. How much water should be supplied to this unit?

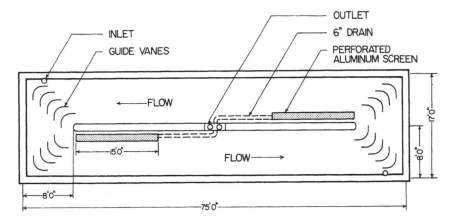

Figure 4.8.   Rectangular circulating pond. (From Burrows, R. F. and Chenoweth,
H. H., *Progressive Fish-Culturist,* 32, 69, 1970. With permission.)

6. Design a circular fish-rearing unit to receive 30 L/sec of water.
7. Calculate the cost to construct the unit above if the walls and floor are 20 cm
   thick and reinforced concrete, poured in place, costs $100/m³.
8. Calculate the difference in cost between the unit you designed above and a
   raceway to receive the same flow.
9. If fish can be reared at 70 kg per L/sec of flow, compare fish densities, in
   kg/m³, between the raceway and the circular pond.
10. The rectangular circulating rearing pond is 75 ft long, 16 ft wide, and 30 in.
    deep. What is the water exchange rate when this unit is supplied with 600 gpm
    of water?

REFERENCES

Burrows, R. E. and H. H. Chenoweth. 1955. *Evaluation of Three Types of Fish Rearing
Ponds.* U.S. Fish and Wildlife Service Research Report 39.

Burrows, R. E. and H. H. Chenoweth. 1970. The rectangular circulating rearing pond.
Progressive Fish-Culturist 32: 67-80.

Buss, K., D. R. Graff, and E. R. Miller. 1970. Trout culture in vertical units. *Progressive
Fish-Culturist* 32: 187-191.

Buss, K. and E. R. Miller. 1971. Considerations for conventional trout hatchery design
and construction in Pennsylvania. *Progressive Fish-Culturist* 33: 86-94.

Haskell, D. C., R. O. Davies, and J. Reckahn. 1960. Factors in hatchery pond design.
New York Fish and Game Journal 7: 113-129.

Larmoyeaux, J. D., R. G. Piper, and H. H. Chenoweth. 1973. Evaluation of circular
tanks for salmonid production. *Progressive Fish-Culturist* 35: 122-131.

Moody, T. M. and R. N. McCleskey. 1978. *Vertical Raceways for Production of
Rainbow Trout.* New Mexico Department of Game and Fish Bulletin No. 17.

Westers, H. and K. M. Pratt. 1977. Rational design of hatcheries for intensive salmonid
culture, based on metabolic characteristics. *Progressive Fish Culturist* 39: 157
165.

CHAPTER 5

The Solubility of Oxygen in Water

THE GAS LAWS

The gas laws, familiar to students of introductory chemistry, are of fundamental importance to fish culture and are therefore reviewed here. The Universal Gas Law, which is the combination of Charles' and Boyle's Laws, relates the volume of a gas to pressure and absolute temperature as follows:

$$\frac{V_1 P_1}{T_1} = \frac{V_2 P_2}{T_2}$$

According to this expression the volume of a gas is proportional to its temperature and inversely proportional to the pressure exerted upon it.

Air is a mixture of gases (Table 5.1). Dalton's Law states that the total pressure of a mixture of gases is equal to the sum of the individual partial pressures of its components. Standard atmospheric pressure or 1.0 atmospheres (atm) is 760 millimeters of mercury (mm Hg). Therefore, the partial pressure of oxygen (PO_2) in a total pressure of 1 atm of dry air is $760 \times 0.20946 = 159.2$ mm Hg. The atmosphere contains water vapor, in addition to the gases listed in Table 5.1, which also exerts a partial pressure called the vapor pressure (Pw). Saturation vapor pressures in relation to air temperature are listed in Table 5.2. The actual Pw is the saturation vapor pressure multiplied by the relative humidity, which is determined from the wet-bulb depression of a sling psychrometer. The PO_2 in moist air at a barometric pressure of P and a known Pw is $P - Pw \times 0.20946$. Hutchinson (1975) suggests that the saturation vapor pressure be subtracted from the barometric pressure when the relative humidity is not known since humidity is more likely to be high than low. For routine aquaculture applications, correction of oxygen tensions for vapor pressure may be omitted except where accurate data are required at warm, humid locations.

Table 5.1. Composition of Dry Air

Gas component	Percent of total volume
Nitrogen	78.084
Oxygen	20.946
Argon	0.934
Carbon Dioxide	0.033
Methane	0.0062
Xenon, Helium, Neon, Oxides of Nitrogen	0.0024

Table 5.2. Saturation Vapor Pressures at Different Air Temperatures

T, °C	P, mm Hg	T, °C	P, mm Hg
0	4.6	18	15.5
1	4.9	19	16.6
2	5.3	20	17.5
3	5.7	21	18.6
4	6.1	22	19.8
5	6.5	23	21.1
6	7.0	24	22.4
7	7.5	25	23.8
8	8.0	26	25.2
9	8.6	27	26.7
10	9.2	28	28.3
11	9.8	29	30.0
12	10.5	30	31.8
13	11.2	31	33.7
14	12.0	32	35.7
15	12.8	33	37.7
16	13.6	34	39.9
17	14.5	35	42.2

Henry's Law describes the solution of atmospheric gases in water. Simply stated, the equilibrium concentration of a gas in water is related to its partial pressure in the atmosphere above the water. The partial pressure of a gas dissolved in water then, is the atmospheric pressure of that gas required to hold it in solution.

CALCULATION OF THE OXYGEN SOLUBILITY IN WATER

Chemists and physicists normally determine the solubility of dissolved gases using the Bunson coefficient, which is 946,000 divided by the Henry's Law constant. These values for any gas and at different temperatures are given in the International Critical Tables. Representative values for the Henry's Law constant for oxygen are given in Table 5.3. To calculate the oxygen content (mg/L) of water in equilibrium with dry air at a pressure of 1 atm, the Bunson coefficient is multiplied by the density of oxygen in mg/L, and then multiplied by 0.20946, the decimal fraction of oxygen in air. For example, the solubility

Table 5.3. Henry's Law Constants (K) for Oxygen

T, °C	K × 10⁷
0	1.933
5	2.205
10	2.486
15	2.766
20	3.044
25	3.344
30	3.599

of oxygen in water at 10°C, in equilibrium with dry air at 1 atm is calculated as follows: the Bunson coefficient divided by the Henry's Law constant is $946,000/2.486 \times 10^7$. The volume of a mole of gas at 10°C is calculated using the Universal Gas Law. The volume of a mole of gas at 0°C (273.15°K) is 22.4 L according to Avogadro's Law. Thus, the volume of a mole of gas at 10°C (283.15°K) is $22.4 \times 283.15/273.15 = 23.2$ L. The density of oxygen then, with a formula weight of 32, at 10°C, is

$$\frac{32000 \text{ mg}}{\text{mol}} \times \frac{\text{mol}}{23.2 \text{ L}} = 1379.3 \text{ mg/L}$$

Thus, the solubility of oxygen with dry air at 10°C is $946,100/2.486 \times 10^7 \times 1379.3 \times 0.20946$, or 10.99 mg/L.

Tables of oxygen solubility have been presented by several authors (Roscoe and Lund 1889; Winkler 1889; Fox 1907, 1909; Truesdale et al. 1955). Mortimer (1956) reviewed these data and concluded that the results of Truesdale et al. (1955) are the most reliable. The data of Truesdale et al. (1955) are recommended by Hutchinson (1975) and Boyd (1979) and are reproduced here (Table 5.4).

Most fish culturists use tables of oxygen solubility when this information is required, but many workers find it more convenient and precise to calculate the oxygen solubility for the conditions at the time and place the value is needed.

Several authors provide empirical formulae for the determination of the equilibrium concentration of oxygen in water (Truesdale et al. 1955; Whipple et al. 1969; Liao 1971). The following regression expression presented by Truesdale et al. (1955) predicts the equilibrium concentration of oxygen in water in the temperature range of 0 to 36°C with a maximum deviation of 0.11 mg/L from their experimentally determined values (Table 5.4):

$$Ce = 14.161 - 0.3943 \text{ T} + 0.0077147 \text{ T}^2 - 0.0000646 \text{ T}^3 \qquad 5.1$$

where Ce is the equilibrium concentration of oxygen in water in mg/L with moist air at a pressure of 1 atm, and T is the temperature in °C. This rather

Table 5.4. **Equilibrium Concentration of Oxygen (Ce)
with Moist Air at a Pressure of 1.0 atm
and Various Temperatures**

T, °C	Ce in mg/L		Deviation mg/L
0	14.16	14.43	0.27
1	13.77	13.95	0.18
2	13.40	13.50	0.10
3	13.05	13.09	0.04
4	12.70	12.71	0.01
5	12.37	12.36	−0.01
6	12.06	12.03	−0.03
7	11.76	11.73	−0.03
8	11.47	11.44	−0.03
9	11.19	11.17	−0.02
10	10.92	10.92	0.00
11	10.67	10.68	0.01
12	10.43	10.45	0.02
13	10.20	10.24	0.04
14	9.98	10.04	0.06
15	9.76	9.85	0.09
16	9.56	9.66	0.10
17	9.37	9.49	0.12
18	9.18	9.32	0.14
19	9.01	9.16	0.15
20	8.84	9.01	0.17
21	8.68	8.86	0.18
22	8.53	8.72	0.19
23	8.38	8.59	0.21
24	8.25	8.46	0.21
25	8.11	8.34	0.23
26	7.99	8.22	0.23
27	7.86	8.10	0.24
28	7.75	7.99	0.24
29	7.64	7.88	0.24
30	7.53	7.78	0.25
31	7.43	7.68	0.25
32	7.33	7.58	0.25
33	7.23	7.49	0.26
34	7.13	7.40	0.27
35	7.04	7.31	0.27

Note: Column 2 was calculated using the equation of Truesdale et al. (1955) and Column 3 using Equation 5.2, modified after Liao (1971). The last column shows the deviation from the values of Truesdale et al. (1955) when the simpler Equation 5.2 is used to calculate Ce.

cumbersome equation can easily be programmed into an electronic pocket calculator. The program will remain in the continuous memory of the small LCD units for at least two years. The temperature at which the oxygen solubility is required is simply entered in the calculator in the compute mode and the value for Ce is automatically calculated.

Liao's (1971) formula for the calculation of Ce is

$$Ce = \frac{132}{T^{0.625}}$$

where T is the temperature in °F. This formula is much easier to use than that of Truesdale et al. (1955) when a programmable calculator is not available, but the values obtained are 4 to 8% higher. By changing the numerator, Liao's (1971) equation becomes

$$Ce = \frac{125.9}{T^{0.625}} \qquad\qquad 5.2$$

and predicts Truesdale's et al. (1955) experimental values with a maximum deviation of 0.27 mg/L (Table 5.4). This is probably sufficiently accurate for fish culture applications.

CORRECTION FOR PRESSURE AND SALINITY

The solubility of a gas in water varies with the atmospheric pressure of that gas in accordance with Henry's Law. Therefore, the calculated value for Ce at standard pressure must be corrected for the atmospheric conditions at the location where the estimate is required. The most accurate way to correct for pressure is to measure the barometric pressure with a barometer when the oxygen solubility is to be calculated. The pressure correction factor, then, is

$$\frac{P}{760} \qquad\qquad 5.3$$

where P = the measured barometric pressure in mm Hg. If a barometer is not available, and for general applications when average values are required, elevation can be used to determine the average atmospheric pressure. Liao (1971) provides the following pressure correction factor,

$$\frac{760}{760 + E/32.8} \qquad\qquad 5.4$$

where E = elevation in feet above sea level. When the elevation is provided in meters above sea level, the pressure correction factor is

$$\frac{1}{1 + E/7600} \qquad\qquad 5.5$$

The calculated oxygen solubility is multiplied by the pressure correction factor to determine the pressure-corrected value.

Dissolved oxygen solubility is also influenced by the concentration of dissolved solids. Truesdale et al. (1955) provide a convenient equation to calculate the depression in Ce caused by salinity:

$$D_s = S(0.0841 - 0.00256\,T + 0.0000374\,T^2)   \qquad 5.6$$

where D_s = reduction in Ce due to salinity, in mg/L, S = salinity in parts per thousand (‰), and T = temperature in °C.

MEASUREMENT OF DISSOLVED OXYGEN CONCENTRATION

Because of fish respiration, water used for aquaculture will seldom contain the equilibrium concentration of dissolved oxygen (DO). The actual DO concentration in water at any time is of fundamental concern to the fish culturist. Therefore, a convenient and accurate means of routinely determining the DO concentration is necessary for the operation of a fish hatchery.

The iodometric method developed by Winkler (1889), incorporating the azide modification to prevent nitrite interference, is widely used in water quality laboratories, but this procedure is too cumbersome for fish hatchery applications. A polarographic DO meter is much more useful and convenient for field applications and the results obtained are as accurate as those from the Winkler method (Boyd 1979). Special care should be taken in purchasing an instrument from a reputable manufacturer and following the instructions carefully with regard to maintenance of the membrane and calibration of the instrument prior to each determination.

EXPRESSION OF DO IN TENSION UNITS

The amount of DO in water is usually expressed in concentration units (mg/L), but for fish respiration considerations it is more usefully expressed as pressure in mm Hg. The oxygen tension in air (PO_2a) is the barometric pressure minus the vapor pressure, multiplied by 0.20946. The tension of a given DO concentration (PO_2w) is the partial pressure of oxygen required to hold that amount of oxygen in solution. Thus, the PO_2w is the decimal fraction of oxygen saturation (measured DO divided by Ce), multiplied by the PO_2a above the water. If a barometer is not available, the average barometric pressure can be adequately estimated from the elevation:

$$P = \frac{577600}{760 + E/32.8} \qquad 5.7$$

where P = average barometric pressure in mm Hg at E feet above sea level, or

$$P = \frac{760}{1 + E/7600} \qquad 5.8$$

where E = meters above sea level.

SAMPLE PROBLEMS

1. The Henry's Law constant for CO_2 at 20°C is 1.070×10^6. Calculate the equilibrium concentration of CO_2 in water at this temperature and standard pressure. CO_2 has a formula weight of 44 g/mol. Note that the Henry's Law constant for CO_2 is much larger than for O_2. The solubility is therefore greater. Why then, is the equilibrium concentration of CO_2 less than for O_2?

2. Calculate the equilibrium concentrations of oxygen in the following waters:
   a) T = 20°C, E = 1500 feet above sea level,
   b) T = 32°C, E = 700 feet below sea level,
   c) T = 50°F, P = 29.9 inches of mercury (in. Hg),
   d) Standard conditions: 20°C, 1 atm pressure.

3. Express the following oxygen concentrations as percent saturation and oxygen tension. Ignore the influence of vapor pressure.
   a) 5.0 mg/L at 50°F when P = 28.92 in. Hg,
   b) 12.0 mg/L at 85°F when E = 600 feet above sea level,
   c) Saturated with oxygen under a pressure of 2.3 atm and a temperature of 10°C, when released to the atmosphere at an elevation of 1500 feet above sea level,
   d) Saturated with oxygen at 10°C, then heated to 20°C when P = 735 mm Hg.

4. Calculate the altitude where half of the atmosphere by weight is above you and half is below you.

5. Calculate the equilibrium concentration of oxygen in a brackish water fish pond at sea level. T = 28°C and S = 17‰.

6. How much oxygen (kg) is in a cubic mile of ocean water whose temperature is 15°C? S = 35‰.

7. Calculate the DO concentration in 25°C water in a sealed plastic bag containing pure oxygen at a pressure of 1 atm.

8. Consider a DO concentration of 5.0 mg/L in 10°C water at an elevation of 500 meters above sea level. If the air temperature is 25°C, how much does the PO_2w change when the relative humidity increases from 10 to 90%?

9. Calculate the percent error introduced by neglecting to correct the calculated PO_2w for Pw in the following examples. Assume that relative humidity is 100%.

48 FLOWING WATER FISH CULTURE

a) Air temperature 0°C, water temperature 8°C, water is saturated with DO, elevation is 350 meters above sea level.
b) Air temperature 90°F, water temperature 50°F, DO is 5.0 mg/L, elevation is 1500 feet above sea level.
c) Air temperature is 30°C, water temperature is 30°C, water is saturated with DO at sea level.
d) Air temperature is 35°C, water temperature is 83°F, DO is 3.5 mg/L, elevation is 1000 feet above sea level.
10. You have established that the lowest DO concentration resulting in acceptable growth of tilapia at 28°C is 3.5 mg/L. What DO concentration would be the same oxygen tension for salmon at 8°C?

REFERENCES

Boyd, C. E. 1979. *Water Quality in Warmwater Fish Ponds*. Auburn University Agricultural Experiment Station, Auburn, AL.

Fox, C. J. J. 1907. On the coefficients of absorption of the atmospheric gases in distilled water and in sea water. *Publ. Circ. Cons. Explor. Mer. 41.*

Fox, C. J. J. 1909. On the coefficients of absorption of nitrogen and oxygen in distilled water and sea water, and of atmospheric carbonic acid in sea water. *Transactions of the Faraday Society* 5: 68-87.

Hutchinson, G. E. 1975. *A Treatise on Limnology: Volume I, Part 2, Chemistry of Lakes*. John Wiley and Sons, New York.

Liao, P. B. 1971. Water requirements of salmonids. *Progressive Fish-Culturist* 33: 210-215.

Mortimer, C. H. 1956. The oxygen content of air saturated fresh waters, and aids in calculating percentage saturation. *Mitt. int. Ver. Limnol.* 6:1-20.

Roscoe, H. E. and J. Lund. 1889. On Schutzenberger's process for the estimation of dissolved oxygen in water. *J. Chem. Soc.* 55: 552-576.

Truesdale, G. A., A. L. Downing, and G. F. Lowden. 1955. The solubility of oxygen in pure water and sea-water. *Journal of Applied Chemistry* 5: 53-62.

Whipple, W., Jr., J. V. Hunter, B. Davidson, F. Dittman, and S. Yu. 1969. *Instream Aeration of Polluted Rivers*. Water Resources Research Institute, Rutgers University, New Brunswick, NJ.

Winkler, L. W. 1889. Die Loslichkeit des sauerstoffs im wasser. *Ber. dtch. Chem. Ges.* 22: 1764-1774.

CHAPTER 6

The Oxygen Requirements of Fish

FISH RESPIRATION PHYSIOLOGY

The fish gill apparatus is a marvelously functional structure for aquatic respiration. Each of four pairs of gill arches in teleosts contains posterior and anterior columns of gill filaments. Each filament bears transverse lamellae of a single cell layer of respiratory epithelium. As water is moved across the gill by the fish's branchial pump, the filaments extend so that their tips touch those of adjacent filaments and the lamellae interdigitate with those on the filaments above and below. In this way, inspired water is forced through very small openings lined with respiratory tissue containing venous blood.

Fish respiratory efficiency is greatly aided by the fact that deoxygenated blood and oxygenated water flow along their contact area in opposite directions. In this way, water containing the most DO first comes in contact with blood partially loaded with oxygen. Farther down the area of blood-water interface, water with a reduced oxygen tension contacts blood whose DO content is also low. Thus, the steepest possible oxygen concentration gradient is maintained across the respiratory epithelium. Carbon dioxide transfer from blood to water is similarly favored by this mechanism.

Water in use for fish husbandry is normally not saturated with DO because of fish respiration. The response of fish to varying degrees of DO reduction is therefore of critical concern to the fish culturist. In respiration fish blood picks up oxygen and releases carbon dioxide at the gills and picks up carbon dioxide and releases oxygen at the tissues. Therefore, the efficiency with which the blood combines with oxygen and carbon dioxide at different tensions determines the reaction of the fish to reduced DO concentrations in the water.

Oxygen dissociation curves (Figure 6.1) depict the extent to which blood is saturated with oxygen at different oxygen tensions. If the difference between the percentage of saturation of the hemoglobin between the tissues and gill is great, much of the oxygen carried by the blood will be released at the tissues

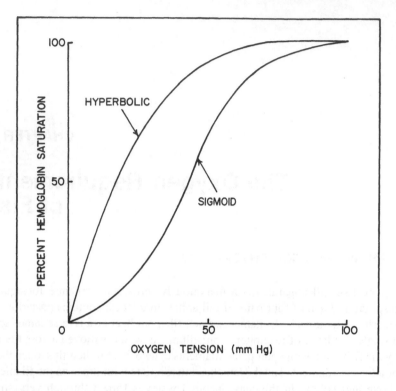

Figure 6.1.   Model hyperbolic and sigmoid fish oxygen dissociation curves.

as illustrated by the sigmoid curve (Figure 6.1) However, little oxygen will be released at the tissues when the differences between the percentages of saturation at oxygen tensions existing at the two locations is small. This situation is depicted by the hyperbolic oxygen dissociation curve (Figure 6.1). The shapes of oxygen dissociation curves are influenced by the species of fish (Fry 1957), temperature (Fry 1957, Cameron 1971) and the tension of carbon dioxide (PCO_2) (Black et al. 1966; Cameron 1971). The tendency is for the oxygen dissociation curve to become hyperbolic (oxygen delivery to the tissues becomes less efficient) with decreasing temperature and decreasing PCO_2.

OXYGEN CONSUMPTION RATES OF CULTURED FISH

The ability to predict the amount of oxygen fish will extract from water under different conditions will allow us to calculate the weight of fish a water stream of known discharge and oxygen content should support. Various authors (Beamish 1964a; Beamish 1964b; Beamish 1964c; Beamish and Dickie 1967; Brett 1962; Brett and Zala 1975; Caulton 1976) have studied the oxygen consumption rates of individual fish in respiration chambers. From this research we know that the amount of DO that fish must extract from water flowing past them depends upon their size, sex, reproductive cycle, activity level, temperature, and oxygen content of the water. Oxygen consumption of confined fish is simpler than in nature because the size of fish within a lot is relatively uniform, the sex ratio is one, cultured fish are generally not allowed to reach sexual maturity, the activity level is constant, and water temperature and oxygen concentrations are constant or fluctuate predictably.

Elliot (1969) studied the oxygen consumption rate of chinook salmon between 1.85 and 17.50 g in weight. He measured the DO concentrations above and below groups of fish under hatchery conditions and developed an empirical method for estimating their rate of oxygen consumption. For fish from 1.85 to 5.90 g,

$$Yn = [0.02420T - 0.7718] - [(0.001242T - 0.04544)(n - 1.85)] \quad 6.1$$

and for fish from 5.90 to 17.50 g,

$$Yn = [0.01917T - 0.5877] - [(0.0003676T - 0.011601)(n - 5.90)] \quad 6.2$$

where Yn = mg/L of DO per pound of fish per gallon per minute (gpm) of flow, n = average individual fish weight in g, and T = temperature in °F.

Liao (1971) collected oxygen consumption data for trout and Pacific salmon in hatcheries and developed an equation that predicts oxygen uptake when water temperature and average fish size are known:

$$O_2 = KT^n W^m \quad 6.3$$

where O_2 = oxygen consumption rate in weight units of DO per 100 weight units of fish per day, K = rate constant, T = water temperature in °F, W = average individual fish weight in lb, and m and n are slopes. The following constants were provided:

Species	T, °F	K	m	n
Salmon	≤50	7.20×10^{-7}	−0.194	3.200
Salmon	>50	4.90×10^{-5}	−0.194	2.120
Trout	≤50	1.90×10^{-6}	−0.138	3.130
Trout	>50	3.05×10^{-4}	−0.138	1.855

Liao's (1971) formula gives nearly identical values as Elliot's (1969) for the oxygen consumption rate of salmon (Figure 6.2) and the two may be used interchangeably for fish smaller than 17.5 g.

Muller-Fuega et al. (1978) conducted an experiment with rainbow trout under hatchery conditions similar to Liao's (1971) investigation. They found a discontinuity in oxygen consumption vs, temperature at 10°C as did Liao (1971). Their expression for predicting the oxygen consumption rate for rainbow trout is,

$$OD = \alpha p^\beta 10^\pi \quad 6.4$$

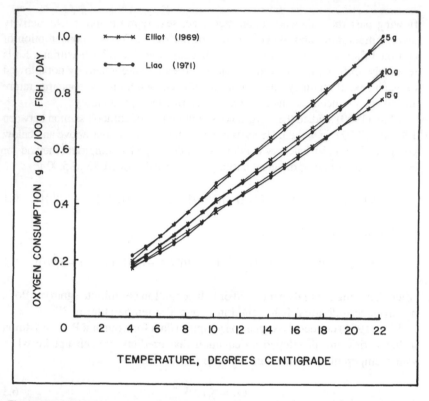

Figure 6.2. Oxygen consumption rates of 5-, 10-, and 15-g Pacific salmon at different water temperatures using the equations of Elliot (1969) and Liao (1971).

where OD = oxygen demand in mg DO/kg of fish per hour, p = weight of individual fish in g, t = temperature in °C, and α, β, and γ are constants. The constant values are as follows:

| | Temperature, °C | |
Constant	4–10	12–22
α	75	249
β	−0.196	−0.142
γ	0.055	0.024

The results obtained from Liao's (1971) and Muller-Fuega's et al. (1978) equations are identical at temperatures greater than 12°C, but different at temperatures less than 12°C (Figures 6.3 to 6.5). This difference is considerable and we shall refer to Willoughby's (1968) extensive examination of hatchery carrying capacity for further information.

Willoughby (1968) related oxygen consumption to the amount of food fed to trout in hatcheries with the following expression:

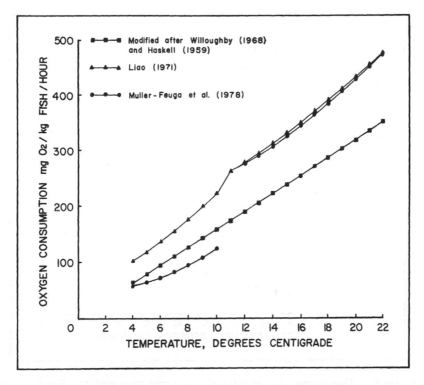

Figure 6.3. Oxygen consumption rates of 50-g trout calculated from Equation 6.5, based on the expressions of Haskell (1959) and Willoughby (1968) and the equations of Liao (1971) and Muller-Fuega et al. (1978).

$$(Oa - Ob) \times 0.0545 \times gpm = \text{pounds of food per day}$$

where Oa = DO in mg/L at the pond influent, Ob = DO in mg/L at the pond effluent, and gpm = pond discharge in gpm for a diet containing 1200 C/lb. When Haskell's (1959) feeding equation is applied to Willoughby's (1968) formula for a given temperature and size of fish, oxygen consumption may be calculated. This is reasonable since both Liao (1971) and Muller-Fuega et al. (1978) collected data on fish being fed at rates based on Haskell's (1959) equation. The two expressions thus combined are

$$Oc = \left(\frac{3 \times C \times \Delta L}{L} \right)(9155.23) \qquad\qquad 6.5$$

where Oc = oxygen consumption in mg DO/kg of fish per hour, C = food conversion, ΔL = daily increase in length, and L = fish length. Data from this equation, using a food conversion of 1.7 and a growth rate of 15 Centigrade temperature units per inch, fall midway between Liao's (1971) and

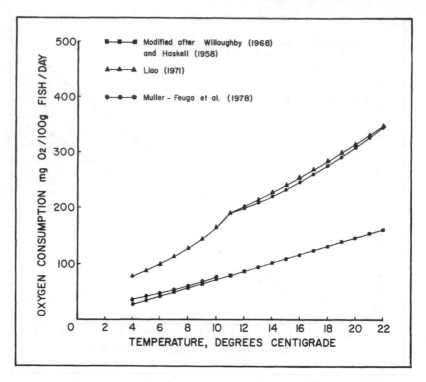

Figure 6.4. Oxygen consumption rates of 250-g trout calculated from Equation 6.5, based on the expressions of Haskell (1959) and Willoughby (1968) and the equations of Liao (1971) and Muller-Fuega et al. (1978).

Muller-Fuega's (1978) results in the lower temperature range and about 25% lower for the higher temperature range (Figure 6.3). Of the three methods available for calculating the oxygen consumption rates of trout in culture, Equation 6.5, based on Haskell's (1959) and Willoughby's (1968) work, may be preferable because it allows for variable growth, food conversion, and diet types to be taken into account.

Andrews and Matsuda (1975) presented oxygen consumption data for channel catfish of different sizes and at several temperatures. Boyd et al. (1978) applied multiple regression analysis to these data to obtain the following expression:

$$\log O_2 = -0.999 - 0.000957W + 0.0000006W^2 + \\ 0.0327T - 0.0000087T^2 + 0.0000003WT$$

6.6

where O_2 = oxygen consumption rate in mg DO/g of fish per hour, W = average individual fish weight in g and T = temperature in °C. The formula can be used to predict the oxygen consumption rate for any size of catfish at any temperature.

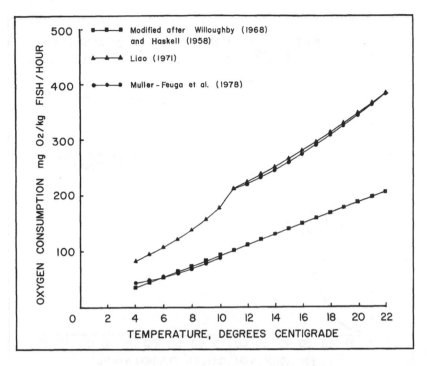

Figure 6.5. Oxygen consumption rates of 500-g trout calculated from Equation 6.5, based on the expressions of Haskell (1959) and Willoughby (1968) and the equations of Liao (1971) and Muller-Fuega et al. (1978).

EFFECTS OF HYPOXIA ON GROWTH

The relationship between the oxygen consumption rate and the environmental oxygen tension has been investigated by Fry and Hart (1948), Graham (1949), Shepherd (1955), Holeton and Randall (1967) and Itazawa (1970). The classical explanation given by Fry (1957) is that oxygen consumption is constant and at a maximum at DO tensions above the incipient limiting level. The incipient lethal level is the tension under which mortality results from prolonged exposure (Figure 6.6). Most workers agree that between the incipient limiting and incipient lethal levels of DO, the oxygen consumption rate is dependent upon environmental DO. This zone of respiratory dependence should also be a zone of reduced growth since less than the maximum amount of oxygen is delivered to the tissues. Holeton and Randall (1967) found that the oxygen consumption rate of rainbow trout did not change over a range of DO tensions of 40 to 160 mm Hg. The fish increased their ventilation volume in order to keep a constant level of oxygen supplied to the tissues.

Whether the mechanism of reduced DO limiting fish growth is decreased oxygen consumption or the osmoregulatory cost of hyperventilation is not particularly germane to practical fish culture. The degree to which growth is

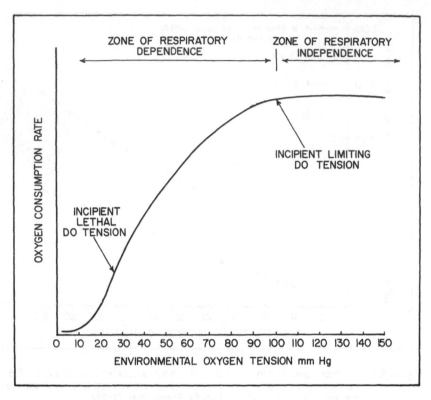

Figure 6.6. Oxygen consumption by fish as related to environmental oxygen tension. [After Fry (1957).]

reduced by various amounts of DO reduction is however, of primary concern because a DO minimum for culture water must be defined that designates at which point the water must be reaerated or discharged.

Brungs (1971), in a carefully controlled long-term study found that the growth of fathead minnows, *Pimephales promelas,* was reduced at all DO concentrations below saturation. Andrews et al. (1973) reported that channel catfish ate less and grew more slowly at 60% saturation (approximately 95 mm Hg) than at 100% saturation. Carlson et al. (1980) found that at 25°C channel catfish growth was reduced at a constant DO exposure of 3.5 mg/L (69 mm Hg), but there was no growth reduction at 5.1 mg/L (100 mm Hg). Larmoyeaux and Piper (1973) reared rainbow trout in a system where the water flowed through a series of troughs, each containing fish. Growth was reduced in the trough where the DO had fallen to 4.2 mg/L (61 mm Hg) by fish respiration upstream, but not in the trough above it where the DO was 4.0 mg/L (71 mm Hg). Ammonia exposure (see Chapter 8) may also have contributed to this reduction in growth. Downey and Klontz (1981) found that rainbow trout growth was not different at 105.3 mm Hg and 138.5 mm Hg, but growth as reduced by 10% at 81.5 mm Hg. Weight gain at 70 mm Hg was 66.6% of that at 128 mm Hg.

ASSIGNMENT OF DO MINIMA FOR HATCHERIES

Willoughby (1968), Piper (1970), Smith and Piper (1975), Leitritz and Lewis (1976) and Westers and Pratt (1977) suggested that aquaculture facilities for trout be designed so that fish are exposed to a minimum DO concentration of 5.0 mg/L. Buss and Miller (1971) called for aeration at trout hatcheries when the DO level was predicted to fall below 5.0 to 7.0 mg/L. Burrows and Combs (1968) recommended a DO minimum of 6.0 mg/L for salmon. Because fish respiration occurs along oxygen tension gradients, criteria for DO minima for fish culture are more usefully and universally expressed in tension units than in concentration units. Furthermore, use of tension units allows changes in oxygen solubility under different conditions to be taken into account.

Reported incipient limiting DO tensions for freshwater fish range from 60 mm Hg for brown bullhead, *Ictalurus nebulosus,* (Grigg 1969) to 100 mm Hg for brook trout (Graham 1949) and rainbow trout (Itazawa 1970). Davis (1975) conducted a statistical analysis of available data on the incipient limiting DO and recommended minimum PO_2 levels of 86 mm Hg for mixed fish populations including salmonids, 73 mm Hg for fish populations not including salmonids, and 90 mm Hg for salmonid populations (Table 6.1). Downey and Klontz (1981) recommended minimum oxygen tensions of 90 mm Hg for trout in hatcheries.

These recommended DO minima may be unnecessarily conservative for practical fish culture. Since the incipient limiting DO is the point where oxygen-dependent growth begins, DO minima based on this value should provide for maximum growth. Perhaps some growth reduction in exchange for the increased carrying capacity that a lower DO minimum would provide is reasonable in aquaculture. As previously cited, several trout hatchery design specialists have recommended a DO minimum of 5.0 mg/L. With 10°C water at an elevation of 1500 meters above sea level, this concentration corresponds to a PO_2 of 73 mm Hg.

Respiratory efficiency is related to the steepness of the oxygen tension gradients between water and blood. Thus, warm-water fish should be able to tolerate lower concentrations of DO than fish in cold water. This is because the tension of a given concentration of oxygen becomes greater as the equilibrium concentration decreases. If a DO minimum of 80 mm Hg for trout is accepted, assigning DO minima of the same PO_2 for other species whose oxygen requirements have not been studied is reasonable. For example, catfish raised at sea level in 28°C water would be exposed to 3.9 mg/L DO when the PO_2 was 80 mm Hg.

Table 6.1. Reported Incipient Limiting DO Tensions
and Recommended DO Minima

Species	Incipient Limiting DO (mm Hg)	Ref.
Brown bullhead	60	Grigg (1969)
Carp	80	Itazawa (1970)
Rainbow trout	78	Irving et al. (1941)
Rainbow trout	100	Itazawa (1970)
Rainbow trout	80	Cameron (1971)
Brook trout	78	Irving et al. (1941)
Brook trout	100	Graham (1949)
Brook trout	78	Irving et al. (1941)
	Recommended DO Minima (mm Hg)	
Fish communities without salmonids	73	Davis (1975)
Fish communities with salmonids	86	Davis (1975)
Salmonid fish communities	90	Davis (1975)

From Davis. (1975).

SAMPLE PROBLEMS

1. Calculate the oxygen consumption rates for fish in the following situations using Liao's formula (Equation 6.3).
    a) Salmon at 8°C, fish weight 102 g each.
    b) Trout at 8°C, each fish is 22 cm long.
    c) Salmon at 52°F, fish weight 42 g each.
    d) Trout at 58°F, each fish is 10 in. long.
2. Use Muller-Fuega's formula (Equation 6.4) for 1b and 1d, convert to common units and compare results.
3. Repeat Problem 1 using Equation 6.5. Convert to common units in order to compare your results with those of Problems 1 and 2.
4. 100 50-g catfish are placed in a concrete tank that is 4 ft long, 2 ft wide and 1 ft deep. The water is 28°C and saturated with DO. How long will it take for the fish to reduce the PO_2 to 50 mm Hg? Use Equation 6.6. Repeat these calculations using Equation 6.5. Compare results.
5. A 10-acre catfish pond 3 ft deep contains 2000 300-g fish per acre. The water temperature is 80°F. An aeration system delivers 1 lb of oxygen per HPh. How large a unit (how many HP) will be required to replace oxygen in the pond removed by the fish? Use Equations 6.5 and 6.6.
6. 500-g catfish are grown in a raceway receiving water geothermally heated to 29°C. What is their oxygen consumption rate? Use Equations 6.5 and 6.6.
7. You have established that 5.0 mg/L is a reasonable DO minimum for trout in 10°C water at 300 meters above sea level. On this basis, what DO minimum would you assign to catfish grown at 27°C and 50 meters above sea level?
8. Compare the amount of DO, in mg/L, available for respiration at the two following locations if influent DO is 100% of saturation and effluent DO is 75 mm Hg.

Location 1: $T = 51°F$, $E = 600$ feet above sea level.
Location 2: $T = 86°F$, $E = 600$ feet above sea level.

9. 500,000 3-in. trout are placed in a raceway that receives 600 gpm of 54°F water. The elevation at the site is 1500 feet above sea level and the oxygen content of the water entering the raceway is 100% of saturation. What will the DO concentration and PO_2 be at the tail of the raceway? Use Equation 6.5.

10. Describe the conflict in fish hatchery management between fish quantity (lb/gpm) and fish quality (healthy fish, able to survive hatchery conditions and resist disease). Based on the literature cited in this chapter, what minimum oxygen level would you assign for fish hatcheries?

REFERENCES

Andrews, J. W., T. Murai, and G. Gibbons. 1973. The influence of dissolved oxygen on the growth of channel catfish. *Transactions of the American Fisheries Society* 102: 835-838.

Andrews, J. W. and Y. Matsuda. 1975. The influence of various culture conditions on the oxygen consumption of channel catfish. *Transactions of the American Fisheries Society* 104: 322-327.

Beamish, F. W. H. 1964a. Seasonal changes in the standard rate of oxygen consumption of fishes. *Canadian Journal of Zoology* 42: 189-194.

Beamish, F. W. H. 1964b. Respiration of fishes with special emphasis on standard oxygen consumption. II. Influence of weight and temperature on respiration of several species. *Canadian Journal of Zoology* 42: 177-188.

Beamish F. W. H. 1964c. Respiration of fishes with special emphasis on standard oxygen consumption. III. Influence of oxygen. *Canadian Journal of Zoology* 42: 355-366.

Beamish, F. W. H. and L. M. Dickie. 1967. Metabolism and biological production in fish. Pp 215-242 in S. D. Gerking (ed.) *The Biological Basis of Freshwater Fish Production.* Blackwell, Oxford, England.

Black, E. C., D. Kirkpatrick, and H. H. Tucker. 1966. Oxygen dissociation curves of the blood of brook trout, *Salvelinus fontinalis,* acclimated to summer and winter temperatures. *Journal of the Fisheries Research Board of Canada* 23: 1-13.

Boyd, C. E., R. P. Romaire, and E. Johnston. 1978. Predicting early morning dissolved oxygen concentrations in channel catfish ponds. *Transactions of the American Fisheries Society* 107: 484-492.

Brett, J. R. 1962. Some considerations in the study of respiratory metabolism in fish, particularly salmon. *Journal of the Fisheries Research Board of Canada* 19: 1025-1038.

Brett, J. R. and C. A. Zala. 1975. Daily pattern of nitrogen excretion and oxygen consumption of sockeye salmon, *Oncorhynchus nerka,* under controlled conditions. *Journal of the Fisheries Research Board of Canada* 32: 2479-2486.

Brungs, W. A. 1971. Chronic effects of low dissolved oxygen concentrations on the fathead minnow, *Pimephales promelas. Journal of the Fisheries Research Board of Canada* 28: 1119-1123.

Burrows, R. E. and B. D. Combs. 1968. Controlled environments for salmon propagation. *Progressive Fish-Culturist* 30: 123-136.

Buss, K. and E. R. Miller. 1971. Considerations for conventional trout hatchery design and construction in Pennsylvania. *Progressive Fish-Culturist* 33: 86-94.

Cameron, J. N. 1971. Oxygen dissociation characteristics of the blood of the rainbow trout, *Salmo gairdneri*. *Comparative Biochemistry and Physiology* 38A: 699-704.

Carlson, A. R., J. Blocher, and L. J. Herman. 1980. Growth and survival of channel catfish and yellow perch exposed to lowered constant and diurnally fluctuating dissolved oxygen concentrations. *Progressive Fish-Culturist* 42: 73-78.

Caulton, M. S. 1976. The effect of temperature on routine metabolism in *Tilapia rendalli boulenger*. *Journal of Fish Biology* 11: 549-553.

Davis, J. C. 1975. Minimal dissolved oxygen requirements of aquatic life with special emphasis on Canadian species: A review. *Journal of the Fisheries Research Board of Canada* 32: 2295-2332.

Downey, P. C. and G. W. Klontz. 1981. *Aquaculture Techniques: Oxygen (PO$_2$) Requirements for Trout Quality*. Idaho Water and Energy Resources Research Institute. University of Idaho, Moscow, ID.

Elliot, J. W. 1969. The oxygen requirements of chinook salmon. *Progressive Fish-Culturist* 31: 67-73.

Fry, F. E. J. 1957. The aquatic respiration of fish. Pp 1-63 in M. E. Brown (ed.) *The Physiology of Fishes*. Volume I. Academic Press, New York, NY.

Fry, F. E. J. and J. S. Hart. 1948. Cruising speed of goldfish in relation to water temperature. *Journal of the Fisheries Research Board of Canada* 7: 169-175.

Graham, J. M. 1949. Some effects of temperature and oxygen pressure on the metabolism and activity of speckled trout, *Salvelinus fontinalis*. *Canadian Journal of Research* 27: 270-288.

Grigg, G. C. 1969. The failure of oxygen transport in a fish at low levels of ambient oxygen. *Comparative Biochemistry and Physiology* 29: 1253-1257.

Haskell, D. C. 1959. Trout growth in hatcheries. *New York Fish and Game Journal* 6: 204-237.

Holeton, G. F. and D. J. Randall. 1967. The effect of hypoxia upon the partial pressure of gasses in the blood and water afferent and efferent to the gills of rainbow trout. *Journal of Experimental Biology* 46: 317-327.

Irving, L., E. C. Black, and V. Stafford. 1941. The influence of temperature upon the combination of oxygen with blood of trout. *Biological Bulletin* 80: 1-17.

Itazawa, Y. 1970. Characteristics of respiration of fish considered from the arterio-venous difference of oxygen content. *Bulletin of the Japanese Society of Scientific Fisheries* 36: 571-577.

Larmoyeaux, J. D. and R. G. Piper. 1973. The effects of water reuse on rainbow trout in hatcheries. *Progressive Fish-Culturist* 35: 2-8.

Leitritz, E. and R. C. Lewis. 1976. *Trout and Salmon Culture — Hatchery Methods*. State of California, Department of Fish and Game, Fish Bulletin 164.

Liao, P. B. 1971. Water requirements of salmonids. *Progressive Fish-Culturist* 33: 210-215.

Muller-Fuega, A., J. Petit, and J. J. Sabaut. 1978. The influence of temperature and wet weight on the oxygen demand of rainbow trout, *Salmo gairdneri* R., in fresh water. *Aquaculture* 14: 355-363.

Piper, R. G. 1970. Know the proper carrying capacities of your farm. *American Fishes and U.S. Trout News* 15: 4-6.

Shepherd, M. P. 1955. Resistance and tolerance of young speckled trout, *Salvelinus fontinalis*, to oxygen lack, with special reference to low oxygen acclimation. *Journal of the Fisheries Research Board of Canada* 12: 387-433.

Smith, C. E. and R. G. Piper. 1975. Lesions associated with chronic exposure to ammonia. Pp 497-514 in W. E. Ribelin and G. Migaki (eds.) *The Pathology of Fishes*. University of Wisconsin Press, Madison, WI.

Westers, H. and K. M. Pratt. 1977. Rational design of hatcheries for intensive salmonid culture, based on metabolic characteristics. *Progressive Fish-Culturist* 39: 157 - 165.

Willoughby, H. 1968. A method for calculating carrying capacities of hatchery troughs and ponds. *Progressive Fish-Culturist* 30: 173-174.

Rearing Density
and Carrying Capacity

INDEPENDENCE OF LOADING RATE
AND REARING DENSITY

Tunison (1945) presented recommended densities for trout in troughs and ponds and demonstrated that acceptable densities increase with increasing fish size. Haskell (1955) related permissible fish density to the feeding rate, recognizing that it was a function of oxygen consumption and metabolite production, and that these parameters were proportional to the amount of food metabolized by the fish.

These authors recommended practical fish densities in weight of fish per volume unit of the fish-rearing container, but it is obvious that permissible fish densities should be related to water flow because the volume of flow determines the amount of oxygen available for respiration and the degree of metabolite dilution. Buss et al. (1970) and Piper (1975) described the independence of water requirements and spatial requirements of cultured fish. It is now conventional terminology to refer to the weight of fish per water flow unit as the loading rate, and the weight of fish per volume unit of rearing space as density. The maximum permissible loading rate is that which results in an effluent DO at the predetermined minimum allowable tension. This is often referred to as carrying capacity. The maximum permissible density may be determined by the hydraulic characteristics of the rearing unit, or by the physical, physiological, or behavioral spatial requirements of the fish.

The important fish husbandry parameters of carrying capacity and maximum density are independent of each other and are presented separately here.

CALCULATION OF HATCHERY CARRYING CAPACITY

Willoughby's Method

Based on Haskell's (1955) principle that oxygen consumption and metabolite production were proportional to the amount of food fed, Willoughby
(1968) presented a formula that related oxygen consumption to the food
ration:

$$(Oa - Ob) \times 0.0545 \times gpm = \text{pounds of food per day}$$

where Oa = DO of incoming water in mg/L, Ob = effluent DO, and gpm =
volume of flow in gal/min. The feed rate in percent of body weight per day is
obtained from Haskell's (1959) formula. Rearranging Willoughby's formula
and inserting the feed rate expression,

$$\text{Loading Rate} = \frac{(Oa - Ob)(0.0545)}{F} \qquad 7.1$$

where loading rate = lb fish/gpm and F = feed rate in lb feed/lb fish per day.
When the value for Ob is the minimum allowable DO, the loading rate is at the
maximum safe level and Equation 7.1 solves for carrying capacity.

Piper's Method

Haskell's (1959) feed rate formula demonstrates that the feed rate is proportional to fish length. Because the permissible weight of fish in a rearing unit is
directly related to the feed rate (Haskell 1959), it follows that carrying capacity
should be related to fish length. Recognizing this, Piper (1975) introduced the
concept of flow index, F = W/L, where F = flow index, W = lb fish/gpm, and
L = fish length in inches. Solving for carrying capacity,

$$\text{Carrying Capacity} = F \times L \qquad 7.2$$

Piper (1975) stated that the desirable flow index for his station at Bozeman,
Montana was 1.5, but recognized that this was a site-specific criterion related
to the DO content of the water. Bruce Cannady (Piper et al. 1982) developed
a table (Table 7.1) that shows recommended flow indices for various conditions of available DO as a function of temperature and elevation, assuming that
influent water is saturated with DO. The values in Cannady's table are based
upon a desirable flow index of 1.5 at 5,000 feet above sea level and 50°F water,
the conditions at the Bozeman station. For example, the recommended flow
index for 48°F water at an elevation of 2,000 feet above sea level is 1.85 (Table
7.1) if influent water is saturated with DO.

Table 7.1. Cannady's Table to Relate Flow Index to Temperature and Elevation

Water temperature °F	Elevation, feet above sea level									
	0	1000	2000	3000	4000	5000	6000	7000	8000	9000
40	2.70	2.61	2.52	2.43	2.34	2.25	2.16	2.09	2.01	1.94
41	2.61	2.52	2.44	2.35	2.26	2.18	2.09	2.02	1.94	1.87
42	2.52	2.44	2.35	2.27	2.18	2.10	2.02	1.95	1.88	1.81
43	2.43	2.35	2.27	2.19	2.11	2.03	1.94	1.88	1.81	1.74
44	2.34	2.26	2.18	2.11	2.03	1.95	1.87	1.81	1.74	1.68
45	2.25	2.18	2.10	2.03	1.95	1.88	1.80	1.74	1.68	1.61
46	2.16	2.09	2.02	1.94	1.87	1.80	1.73	1.67	1.61	1.55
47	2.07	2.00	1.93	1.86	1.79	1.73	1.66	1.60	1.54	1.48
48	1.98	1.91	1.85	1.78	1.72	1.65	1.58	1.53	1.47	1.42
49	1.89	1.83	1.76	1.70	1.64	1.58	1.51	1.46	1.41	1.36
50	1.80	1.74	1.68	1.62	1.56	1.50	1.44	1.39	1.34	1.29
51	1.73	1.67	1.62	1.56	1.50	1.44	1.38	1.34	1.29	1.24
52	1.67	1.61	1.56	1.50	1.44	1.39	1.33	1.29	1.24	1.19
53	1.61	1.55	1.50	1.45	1.39	1.34	1.29	1.24	1.20	1.15
54	1.55	1.50	1.45	1.40	1.34	1.29	1.24	1.20	1.16	1.11
55	1.50	1.45	1.40	1.35	1.30	1.25	1.20	1.16	1.12	1.07
56	1.45	1.40	1.35	1.31	1.26	1.21	1.16	1.12	1.08	1.04
57	1.41	1.36	1.31	1.27	1.22	1.17	1.13	1.09	1.05	1.01
58	1.36	1.32	1.27	1.23	1.18	1.14	1.09	1.05	1.02	0.98
59	1.32	1.28	1.24	1.19	1.15	1.10	1.06	1.02	0.99	0.95
60	1.29	1.24	1.20	1.16	1.11	1.07	1.03	0.99	0.96	0.92
61	1.25	1.21	1.17	1.13	1.08	1.04	1.00	0.97	0.93	0.90
62	1.22	1.18	1.14	1.09	1.05	1.01	0.97	0.94	0.91	0.87
63	1.18	1.14	1.11	1.07	1.03	0.99	0.95	0.92	0.88	0.85
64	1.15	1.12	1.08	1.04	1.00	0.96	0.92	0.89	0.86	0.83

From Piper et al. (1982).

If the water supplied to a rearing unit is not saturated with DO, such as in the case of raceways in series, Piper et al. (1982) recommend decreasing the flow index in proportion to the decrease in available DO. For example, a site with an equilibrium concentration of DO of 10 mg/L and a desired DO minimum of 4.7 mg/L has $10 - 4.7 = 5.3$ mg/L of DO available for fish respiration. If the influent DO is 8 mg/L, $8 - 4.7 = 3.3$ mg/L is available, and the selected flow index from Table 7.1 should be multiplied by 3.3/4.7, or 0.70.

Calculation of Carrying Capacity from Oxygen Consumption

Knowledge of the oxygen consumption rate of fish allow direct calculation of their water requirement. Empirical formulae for oxygen consumption rates are available for chinook salmon (Elliot 1969; Liao 1971), trout (Willoughby 1968; Liao 1971; Muller-Fuega et al. 1978), and catfish (Boyd et al. 1978) (Chapter 6). Derivation of the required units conversions follow. Let CC = carrying capacity in lb/gpm, and $C_i - C_m$ = the DO available for respiration in mg/L. For chinook salmon (Elliot 1969),

$$CC = (C_i - C_m)\ \text{mg/L} \times \frac{\text{lb fish}}{Y_n\ \text{mg/L} \times \text{gpm}}$$

and

$$CC = \frac{Ci - Cm}{Yn} \qquad 7.3$$

where Yn = the oxygen consumption rate in (mg/L)/(lb fish/gpm) calculated from Equation 6.1 or 6.2. For Pacific salmon and trout (Liao 1971),

$$CC = \frac{(Ci - Cm)\,mg}{L} \times \frac{100\,lb\,fish/day}{Oc\,lb} \times \frac{lb}{453600\,mg} \times \frac{3.78\,L}{gal} \times \frac{1440\,min}{day}$$

and

$$CC = \frac{1.2\,(Ci - Cm)}{Oc} \qquad 7.4$$

where Oc = the oxygen consumption rate in (lb DO)/(100 lb fish/day) calculated from Equation 6.3. For trout (Muller-Fuega et al. 1978),

$$CC = \frac{(Ci - Cm)\,mg}{L} \times \frac{kg\,fish/hr}{OD\,mg} \times \frac{2.2\,lb}{kg} \times \frac{60\,min}{hr} \times \frac{3.78\,L}{gal}$$

and

$$CC = \frac{499\,(Ci - Cm)}{OD} \qquad 7.5$$

where OD = the oxygen consumption rate in (mg)/(kg fish/hr) calculated from Equation 6.4. For catfish (Boyd et al. 1978),

$$CC = \frac{(Ci - Cm)\,mg}{L} \times \frac{g\,fish/hr}{O_2\,mg} \times \frac{lb}{453.6\,g} \times \frac{60\,min}{hr} \times \frac{3.78\,L}{gal}$$

and

$$CC = \frac{0.5\,(Ci - Cm)}{O_2} \qquad 7.6$$

where O_2 = the oxygen consumption rate in (mg)/(g fish/hr) calculated from Equation 6.5.

Comparison of Methods for Calculation of Carrying Capacity

Direct calculation of carrying capacity is necessary for species other than trout and Pacific salmon because indirect methods have not been described for them. Three authors (Willoughby 1968; Liao 1971; Muller-Fuega et al. 1978) provide methods for determining the oxygen consumption rates of trout and Elliot (1969) and Liao (1971) describe similar procedures for Pacific salmon (Chapter 6). All methods give slightly different results, but Willoughby's (1968) method may be preferable because it allows for variable growth, food conversion, and diet composition to be taken into account. Piper's (1975) flow index method is much simpler to use than calculating carrying capacity from the oxygen consumption rate, but gives substantially lower values (Figures 7.1 and 7.2).

FISH REARING DENSITY

From the preceding sections it is evident that the water requirements of cultured fish are fairly well known. Their spatial requirements, however, are rather poorly defined. A fish-rearing unit is designed to receive a particular volume of water flow. Thus the hydraulic requirements of the rearing container set the maximum fish density because carrying capacity must not be exceeded. Westers (1970) provided a series of graphs that relate fish density to the water exchange rate of the rearing container. Since the exchange rate is determined by water flow, the following expression (Westers 1970) relates carrying capacity to fish density:

$$lb/ft^3 = \frac{lb/gpm \times R}{8} \qquad 7.7$$

where R = the number of water exchanges per hour. This useful formula relates fish density and loading rate according to the hydraulic characteristics of the rearing unit, but does not address the spatial requirements of the fish.

Spatial Requirements of Cultured Fish

The behavioral aspects of fish spatial requirements can be addressed by the example in which individuals of an aggressive species such as bluegill are successively added to an aquarium. Adverse effects are not noted until the fifth fish is added because there are only four distinct territories (corners) to defend in the aquarium. As more individuals are added, this territoriality breaks down and behavior no longer limits the density at which the fish may be held. Even

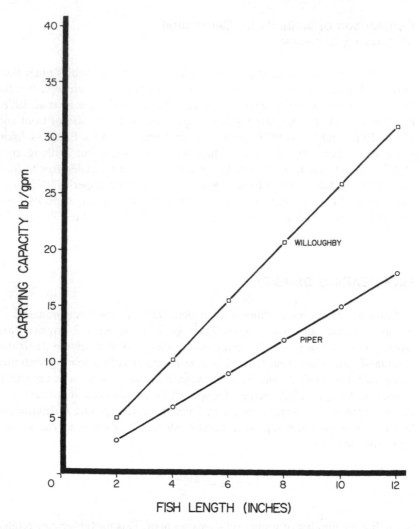

Figure 7.1. Trout carrying capacity as a function of fish length calculated from Willoughby (1968) and Piper (1975). Assumptions used are flow index = 1.5, elevation = 5000 feet above sea level, influent water saturated with DO, DO minimum = 75 mm Hg.

low hatchery rearing densities are great enough to preclude behavioral limitations, with the possible exception of Atlantic salmon.

Some references infer that crowding increases the incidence of infectious disease and fin erosion. Soderberg and Krise (1986) reported greater mortality due to bacterial disease in lake trout reared at 15.4 lb/ft^3 than in those reared at lower densities, up to 9.2 lb/ft^3. Their experimental rearing containers were constructed from rough plastic netting that may have caused abrasions allowing epizootics to occur in the high density treatment. Buss et al. (1970) reared

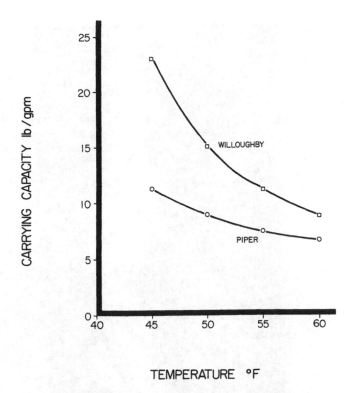

Figure 7.2. Carrying capacity of 6-in. trout at different temperatures calculated from Willoughby (1968) and Piper (1975). Assumptions: Piper — flow index is for 5000 feet above sea level from Table 6.1; Willoughby — growth rate = 7.1 MTU/cm, C = 1.5, elevation = 5000 feet above sea level, influent water saturated with DO, DO minimum = 75 mm Hg.

trout at 34 lb/ft³ in smooth-sided containers without disease incidence (Figure 7.3). A correlation between fin erosion and rearing density in Atlantic salmon has been reported by Westers and Copeland (1973) and Maheshkumar (1985), but such an effect was not noted in other studies with lake trout (Soderberg and Krise 1987) and Atlantic salmon (Schneider and Nicholson 1980; Soderberg and Meade 1987; Soderberg et al. 1993a).

Density Index

Piper (1975) related recommended fish density to fish length in the term "density index". The density index is simply the allowable density in lb/ft³ divided by the fish length in inches. A density index of 0.5 was recommended for trout production in hatcheries (Piper 1975). Maximum rearing density is calculated as

$$Rearing\ Density = L \times 0.5 \qquad 7.8$$

Figure 7.3. Trout reared at extreme densities in a hatching jar. Spatial considerations did not affect fish performance as long as sufficient water flow for respiration was maintained. Photo courtesy of Keen Buss.

Thus, 8-in. trout could safely be reared at 4 lb/ft^3 as long as their water requirements for respiration and metabolite dilution were met.

Density index has been widely adopted for salmonid hatchery design, but trout have been reared at densities considerably greater than are common at most hatcheries. Clary (1979) found that trout were not adversely affected when reared at a density index of 0.8. Soderberg and Krise (1986) noted no effects on 7.7-in. lake trout reared at a density index of 1.0 (7.7 lb/ft^3). In a similar study (Soderberg et al. 1987), 4.7-in. lake trout were successfully reared at density indices as high as 2.0 (9.4 lb/ft^3). These results indicate that allowable fish densities may not be related to fish length and that trout may be safely reared at densities of around 10 lb/ft^3 regardless of size. Because small fish have greater water requirements than larger fish, their high-density culture requires greater flows than equal densities of larger fish.

Atlantic salmon appear to have different spatial requirements than other salmonid species. Keenleyside and Yamamoto (1962) reported that Atlantic salmon must maintain discrete feeding territories for optimum growth and survival. Fenderson and Carpenter (1971) observed that Atlantic salmon hatchery production is related to their weight per unit of area of the rearing container rather than its volume, because they orient to a surface rather than occupying the entire water column. Early recommendations on rearing densities for Atlantic salmon range from 0.75 lb/ft^2 of rearing unit floor surface (Piggins 1971) to 1.5 lb/ft^2 (Peterson et al. 1972). However Westers and Copeland (1973) found that growth did not decrease until densities approached 3.0 lb/ft^2. Soderberg and Meade (1987) reared Atlantic salmon at densities up to 4.29 lb/ft^2 in 8.0°C water and up to 2.86 lb/ft^2 at 17.5°C (Soderberg et al. 1993b) without observing adverse effects on growth or survival.

Schmittou (1969) successfully reared channel catfish in cages suspended in ponds at densities as high as 421 lb/m^3 (11.9 lb/ft^3). Fish growth, feed conversion, and survival were not affected by density in the range from 254 to 418 lb/m^3 (7.2 to 11.8 lb/ft^3). Thus a maximum density for catfish in cages was not demonstrated (Schmittou 1969). Ray (1979) determined that catfish could be reared in raceways at a density of at least 10 lb/ft^3. Hickling (1971) reported that common carp are reared in Japan at densities as high as 278 kg/m^3 (17.4 lb/ft^3).

The factors that determine practical densities for intensive fish culture are not known, but the evidence presented above indicates that crowding alone does not limit salmonid production at rearing densities 2 to 3 times greater than are usually experienced in hatchery conditions. The consequence of this conclusion to salmonid hatchery design is that most facilities provide more rearing space than is necessary for their production requirements. Rearing space at existing hatcheries could be reduced to decrease production costs, and rearing units at proposed hatcheries can be made smaller than those typically used for salmonid production.

SAMPLE PROBLEMS

1. Three trout raceways in series are supplied with water that is saturated with DO. The influent DO to the first unit is the equilibrium concentration. Use Piper's (Equation 7.2) formula and the equation modified from Willoughby's method (Equation 7.1) to calculate carrying capacity of each of the three units assuming that aeration between units restores DO to 80% of saturation.

2. How many pounds of 400-g catfish could you raise in a 1000-cfs discharge from a power plant that averages 85°F? The minimum allowable DO tension is 75 mm Hg. What pond volume would be required if the maximum rearing density is 10 lb/ft^3?

3. A trout hatchery in Pennsylvania has a flow of 3000 gpm at 47°F. How many 8-in. trout could be held there at one time? How many feet of raceways are required if they are 6 ft wide and 2 ft deep?

4. What is the carrying capacity of 10-in. trout at a North Carolina trout farm which receives 400 gpm of 58°F water at an elevation of 2100 feet above sea level. The water is used 16 times with reaeration to 90% of saturation between uses.

5. Brackish groundwater with at a temperature of 28°C and a salinity of 10 ‰ is supplied to a tilapia farm on the coast of Kenya. How much water is required to grow 100,000 kg of 450-g fish? Assume that the oxygen consumption rate for tilapia is the same as for catfish.

6. Prepare a graph that compares the results of Equations 7.1 and 7.2 for calculating trout carrying capacity to direct calculation from the oxygen consumption rate using Liao's (Equation 7.4) or Muller-Fuega's (Equation 7.5) formula. How are they similar? How do they differ?

7. Using the data obtained in Problem 6, calculate the permissible density from Wester's (Equation 7.7) formula, using an exchange rate, R, of four per hour. How do these values compare to Piper's recommendation of a density index of 0.5?

8. How many 6-in. salmon smolts could be reared in a 450-gpm water supply at sea level that averaged 51°F? Assume one water use.

9. Calculate the difference in carrying capacity of trout fed a 1200 C/lb diet and a 1500 C/lb diet.

10. How many 6-in. catfish should be stocked in a cylindrical cage 1 m in diameter and 1.2 m deep if their density is to be 200 kg/m³ when they reach a harvest size of 14 in.?

REFERENCES

Boyd, C. E., R. P. Romaire, and E. Johnston. 1978. Predicting early morning dissolved oxygen concentrations in channel catfish ponds. *Transactions of the American Fisheries Society* 107: 484-492.

Buss, K., D. R. Graff, and E. R. Miller. 1970. Trout culture in vertical units. *Progressive Fish-Culturist* 32: 187-191.

Clary, J. R. 1979. High density trout culture. *Salmonid* 2: 8-9.

Elliot, J. W. 1969. The oxygen requirements of chinook salmon. *Progressive Fish-Culturist* 31: 67-73.

Fenderson, O. E. and M. R. Carpenter. 1971. Effects of crowding on the behavior of juvenile hatchery and wild landlocked Atlantic salmon *(Salmo salar)*. *Animal Behavior* 19: 439-447.

Haskell, D. C. 1955. Weight of fish per cubic foot of water in hatchery troughs and ponds. *Progressive Fish-Culturist* 17: 117-118.

Haskell, D. C. 1959. Trout growth in hatcheries. *New York Fish and Game Journal* 6: 205-237.

Hickling, C. F. 1971. *Fish Culture*. Faber and Faber. London, England.

Keenleyside, M. H. A. and F. T. Yamamoto. 1962. Territorial behavior of juvenile Atlantic salmon *(Salmo salar L.)*. *Behavior* 19: 139-169.

Liao, P. B. 1971. Water requirements of salmonids. *Progressive Fish-Culturist* 33: 210-215.

Maheshkumar, S. 1985. The epizootiology of finrot in hatchery-reared Atlantic salmon *(Salmo salar)*. Master's thesis. University of Maine, Orono.

Muller-Fuega, A., J. Petit, and J. J. Sabaut. 1978. The influence of temperature and wet weight on the oxygen demand of rainbow trout, *Salmo gairdneri* R., in fresh water. *Aquaculture* 14: 355-363.

Peterson, H. H., O. T. Carlson, and S. Johansson. 1972. *The rearing of Atlantic salmon.* Astro-Ewos AB, Sodertalje, Sweden.

Piggins, D. J. 1971. Smolt rearing, tagging and recapture techniques in a natural river system. *International Atlantic Salmon Foundation Special Publication Series* 2: 63-82.

Piper, R. G. 1975. *A Review of Carrying Capacity Calculations for Fish Hatchery Rearing Units.* U.S. Fish and Wildlife Service Fish Culture Development Center (Bozeman, Montana) Information Leaflet No. 1.

Piper, R. G., I. B. McElwain, L. E. Orme, J. P. McCraren, L. G. Fowler, and J. R. Leonard. 1982. *Fish Hatchery Management.* U.S. Department of the Interior, Fish and Wildlife Service, Washington, D.C.

Ray, L. 1979. Channel catfish production in geothermal water. Pp 192-195 in L. J. Allen and E. C. Kinney (eds.) *Proceedings of the Bio-Engineering Symposium for Fish Culture.* Publication 1, Fish Culture Section, American Fisheries Society, Bethesda, MD.

Schmittou, H. R. 1969. Developments in the culture of channel catfish, *Ictalurus punctatus* R., in cages suspended in ponds. 23rd Annual Conference of the Southeastern Association of Game and Fish Commissioners, Mobile, AL.

Schneider, R. and B. L. Nicholson. 1980. Bacteria associated with finrot disease in hatchery-reared Atlantic salmon *(Salmo salar)*. *Canadian Journal of Fisheries and Aquatic Sciences* 37: 1505-1513.

Soderberg, R. W. and W. F. Krise. 1986. Effects of density on growth and survival of lake trout, *Salvelinus namaycush*. *Progressive Fish-Culturist* 48: 30-32.

Soderberg, R. W. and J. W. Meade. 1987. Effects of rearing density on the growth, survival and fin condition of Atlantic salmon. *Progressive Fish-Culturist* 49: 280-283.

Soderberg, R. W., D. S. Baxter, and W. F. Krise. 1987. Growth and survival of fingerling lake trout reared at four different densities. *Progressive Fish-Culturist* 49: 284-285.

Soderberg, R. W. and W. F. Krise. 1987. Fin condition of lake trout, *Salvelinus namaycush* Walbaum, reared at different densities. *Journal of Fish Diseases* 10: 233-235.

Soderberg, R. W., J. W. Meade, and L. A. Redell. 1993a. Fin condition of Atlantic salmon reared at high densities in heated water. *Journal of Aquatic Animal Health* 5: 77-79.

Soderberg, R. W., J. W. Meade, and L. A. Redell. 1993b. Growth, survival and food conversion of Atlantic salmon reared at four different densities with common water quality. *Progressive Fish-Culturist* 55: 29-31.

Tunison, A. V. 1945. *Trout Feeds and Feeding.* Cortland Experimental Hatchery, Cortland, New York. mimeo.

Westers, H. 1970. Carrying capacity of salmonid hatcheries. *Progressive Fish-Culturist* 32: 43-46.

Westers, H. and J. Copeland. 1973. *Atlantic salmon rearing in Michigan.* Michigan Department of Natural Resources, Fisheries Division, Technical Report 73-27, Lansing.

Willoughby, H. 1968. A method for calculating carrying capacities of hatchery troughs and ponds. *Progressive Fish-Culturist* 30: 173-174.

CHAPTER 8

Reaeration of Flowing Water

When fish respiration has reduced the DO tension to the minimum allowable level, water must be reconditioned by aeration to be of further use for fish production. The extent to which water may be reused with aeration as the only treatment measure depends upon the accumulation of toxic metabolic by-products. A detailed description of this fish husbandry parameter follows in Chapter 9.

AERATION THEORY

Transfer of oxygen into water is a three-stage process in which gaseous oxygen is transferred to the surface film, diffuses through the surface film, and finally moves into the liquid bulk by convection. Since oxygen enters water by diffusion, the rate of oxygen transfer depends upon the area of air-water interface and the oxygen deficit of the water. Diffusion of atmospheric oxygen in aquaculture systems where water is quiescent or moving in laminar flow, and oxygen deficits are quite small is too slow to be an important source of DO for fish respiration unless the area of air-water interface is increased by turbulence or agitation. Aeration of water streams used for aquaculture can be accomplished by gravity where the energy released when water loses altitude is used to increase the air-water interface, or by mechanical devices that spray water into the air or inject air or oxygen into the water.

GRAVITY AERATION DEVICES

The most logical means of improving oxygen regimes of cultured fish is by gravity fall of water between production units, which is provided by the topography at the facility (Figures 8.1 and 8.2). The extent to which water is reaerated by gravity is of fundamental concern for practical fish culture in flowing water.

Haskell et al. (1960) compared aeration by water passage over a simple weir (Figures 8.3A and 8.4) with that obtained by water flow over a splashboard

Figure 8.1. Raceways on a level site provide little elevational difference between units.

(Figures 8.3B and 8.5) that broke the water fall partway down, and that from flow over various screens and slat arrangements at the dam. Chesness and Stephens (1971) evaluated several devices for increasing oxygen transfer over a gravity fall including a splashboard, an inclined sheet of corrugated roofing material (Figures 8.3C and 8.6), a similar corrugated sheet pierced with holes (Figures 8.3D and 8.7), and an open stairstep device referred to as a lattice (Figure 8.3E). Tebbutt (1972) studied aeration down closed stairstep arrangements called cascades (Figure 8.3F).

The following equation (Downing and Truesdale 1955) can be used to evaluate and compare aeration devices:

$$E = 100 \times \frac{\text{actual increase in DO}}{\text{possible increase in DO}}$$

or

$$E = 100 \times \frac{Cb - Ca}{Ce - Ca} \qquad 8.1$$

where E = efficiency, Cb = DO in mg/L below the device, Ca = DO in mg/L above the device, and Ce = equilibrium concentration of DO. Selected data on measured efficiencies of some gravity aerators over various distances of water fall are presented in Table 8.1.

Figure 8.2. Two views of an aquaculture site where topography allows for considerable gravity aeration between production units.

The efficiency equation can be rearranged to solve for the expected DO below an aeration device of known efficiency,

$$Cb = \frac{E\,(Ce - Ca)}{100} + Ca$$

8.2

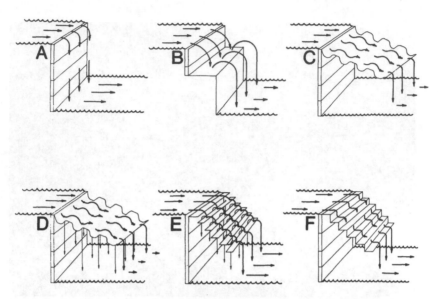

Figure 8.3. Diagrams of gravity aerators. A, Simple weir (Haskell et al. 1960; Chesness and Stephens 1971); B, splashboard (Haskell et al. 1960; Chesness and Stephens 1971); C, inclined corrugated sheet (Chesness and Stephens 1971); D, inclined corrugated sheet with holes (Chesness and Stephens 1971); E, lattice aerator (Chesness and Stephens 1971); and F, cascade aerator (Tebbutt 1972). (From Soderberg, R. W., *Progressive Fish-Culturist,* 44, 91, 1982. With permission.)

Figure 8.4. Water flowing from one raceway to another over a simple weir.

Figure 8.5. Gravity aeration accomplished with splashboards placed between ponds.

Figure 8.6. Inclined sheet of corrugated roofing material used to increase aeration capacity of a gravity fall.

Figure 8.7. Gravity aerator built with corrugated roofing material pierced with holes. Photo courtesy of R. O. Smitherman.

Table 8.1. Selected Data on Measured Efficiencies of Some Gravity Aerators over Various Distances of Water Fall

Device	Water fall (cm)	Efficiency (%)
Simple Weir	22.9[1]	6.2
	30.5[2]	9.3
	61.0[2]	12.4
Inclined Corrugated Sheet[2]	30.5	25.3
	61.0	43.0
Inclined Corrugated Sheet With Holes[2]	30.5	30.1
	61.0	50.1
Splashboard	22.9[1]	14.1
	30.5[2]	24.1
	61.0[2]	38.1
Lattice[2]	30.5	34.0
	61.0	56.2
Cascade[3]	25.0	23.0
	50.0	33.4
	75.0	41.2
	100.0	52.4

[1] Haskell et al. (1960)
[2] Chesness and Stephens (1971)
[3] Tebbutt (1972)
From Soderberg, R. W., *Progressive Fish-Culturist,* 44, 91, 1982. With permission.

Suppose there is a 12-in. drop between two ponds with a simple weir separating them. If the water temperature is 10°C and the elevation is 183 meters above sea level, the solubility of DO is 10.67 mg/L (Chapter 4). If the fish loading in the upstream pond is such that the DO concentration is depressed to 5 mg/L, Ca will be 5.0. We know that a 12-in. fall over a simple weir is 9.3% efficient (Table 8.1). The DO below the weir, then, is

$$Cb = \frac{9.3\,(10.67 - 5.0)}{100} + 5.0 = 5.53 \text{ mg/L}$$

Notice that this procedure compensates for the oxygen deficit of the water being aerated, and for a given amount of increase in the air-water interface provided by a particular aeration device, the actual oxygen transfer is related to the deficit Ce − Ca. In the previous example, 5.53 − 5.0, or 0.53 mg/L of DO was added to the water. If this water is sent over the aeration device a second time,

$$Cb = \frac{9.3\,(10.67 - 5.53)}{100} + 5.53 = 6.01 \text{ mg/L}$$

and the oxygen transferred is 6.01 − 5.53, or 0.48 mg/L, because the oxygen deficit of the water being aerated was lower.

Although the benefit of gravity aeration device can be considerable, most aquaculture sites require mechanical aeration to realize the production potential of their water supplies.

MECHANICAL AERATION DEVICES

Mechanical units that agitate the water surface are commonly used in flowing water aquaculture systems because of their convenience and ease of installation. Aerators are evaluated and compared on the basis of their ability to transfer oxygen to water. Tests are conducted under standard conditions of 760 mm Hg pressure, 20°C temperature, and zero DO in the water being aerated. The amount of oxygen added to the water in a given amount of time under a certain power level is measured. The rating in kilograms of oxygen per shaft kilowatt per hour (kg/kWh) or pounds of oxygen per horsepower per hour (lb/HPh) is given by the aerator manufacturer as a measure of its efficiency and can be used to compare units. Actual oxygen transfer depends upon the oxygen deficit because as saturation is approached, an increasing amount of power is required per unit of DO transferred. Reaeration above 95% of saturation can seldom be justified on a cost basis (Mayo 1979). Westers and Pratt (1977) list 90% of saturation as a reasonable design criterion for reaerated water. Since aquaculture systems operate at the relatively high DO minima of 3 to 7 mg/L, actual transfer rates will be less than those determined under standard conditions.

Surface aerators are generally rated to transfer 1.9 to 2.3 kg/kWh under standard conditions (Eckenfelder 1970). Whipple et al. (1969) found that mechanical aerators in polluted rivers generally provided oxygen transfer rates of 0.61 kg/kWh or less, but their test water was higher in oxygen demand than is usual for aquaculture effluents. Soderberg et al. (1983) reported an average transfer rate of 0.83 kg/kWh in static-water trout ponds where fish were heavily fed and aeration began when DO tensions reached 75 mm Hg. Aeration of flowing water should be more efficient than in static pools because processed water is continually being replaced from upstream rather than being recirculated around the unit. To estimate aeration requirements, a conservative value such as 0.6 kg/kWh may be used or the oxygen transfer may be estimated from

$$RT = RS \frac{\left(\beta\, Ce_t - Ca\right)\left(1.025^{T-20}\right)(\alpha)}{Ce_{20}}$$

where RT = actual oxygen transfer, RS = oxygen transfer under standard conditions, Ce_t = oxygen solubility at the aeration site, Ca = DO concentration of the water to be aerated, T = temperature in °C, Ce_{20} = oxygen solubility under standard conditions, ß = DO solubility in aerated water/DO solubility in clean water, and α = DO transfer in aerated water/DO transfer in clean water. Since aquaculture waters are generally relatively unpolluted, values of 0.85 and 1.0 may be used for α and ß, respectively. The solubility of DO at 20°C and 760 mm Hg pressure is 8.84 mg/L (Chapter 4), thus the oxygen transfer equation reduces to

$$RT = RS \frac{\left(Ce_t - Ca\right)\left(1.025^{T-20}\right)(0.85)}{8.84} \qquad 8.3$$

The following example illustrates the use of this formula. Suppose an aerator is rated by the manufacturer to transfer oxygen at 2.0 kg/kWh under standard conditions. The actual oxygen transfer at a site where the water temperature is 10°C, barometric pressure is 725 mm Hg, and the DO of the water to be aerated is 5.0 mg/L, may be estimated as follows:

$$RS = 2.0 \text{ kg/kWh}$$
$$Ce_t = 10.42 \text{ mg/L (Chapter 4)}$$
$$Ca = 5.0 \text{ mg/L}$$
$$T = 10°C$$
$$RT = 2.0 \frac{(10.42 - 5.0)\left(1.025^{10-20}\right)(0.85)}{8.84} = 0.81 \text{ kg/kWh}$$

When the actual oxygen transfer rate, RT, has been estimated, the aeration capability of a particular unit and set of conditions can readily be obtained. For the above example, suppose that a 1.0-kW unit will be placed in a water flow of 4000 L/min. The DO concentration below the aerator (Cb) is

$$Cb = Ca + \frac{0.81 \text{ kg}}{\text{kWh}} \times \frac{10^6 \text{ mg}}{\text{kg}} \times 1.0 \text{ kW} \times \frac{\text{min}}{4000 \text{ L}} \times \frac{\text{hr}}{60 \text{ min}}$$

and,

$$Cb = 5.0 + 3.38 = 8.38 \text{ mg/L}$$

Similarly, an aerator may be sized for a particular job. Suppose that for the above example, a unit that will return the DO to 90% of saturation is desired. The amount of oxygen that would be required is

$$0.9(10.42) - 5.0 = 4.38 \text{ mg/L} \times \frac{\text{kg}}{10^6 \text{ mg}} \times \frac{4000 \text{ L}}{\text{min}} \times \frac{60 \text{ min}}{\text{hr}} = \frac{1.05 \text{ kg}}{\text{hr}}$$

The size of the unit required in kW of shaft power is

$$\frac{1.05 \text{ kg}}{\text{hr}} \times \frac{\text{kWh}}{0.81 \text{ kg}} = 1.3 \text{ kW}$$

Burrows and Combs (1968) describe an aeration system appropriate for aerating an entire hatchery water supply (Figure 8.8).

DIFFUSER AERATION DEVICES

Diffused air systems introduce air or oxygen into the water. The effectiveness of oxygen transfer is obviously a function of bubble size, as it affects the area of air-water interface, the oxygen tension in the bubble, as it affects the oxygen deficit, and the travel time of the bubble in the water column. The air-water interface may be increased by decreasing the bubble size, and the oxygen concentration gradient between the bubble and the water can be increased by using pure oxygen or increasing the pressure of the injected gas.

Colt and Tchobanoglous (1979) list oxygen transfer rates for diffused air systems under standard conditions at 0.6 to 2.0 kg/kWh, depending upon bubble size. The distance of travel of the bubble in the water column is not given.

When correcting standard oxygen transfer rates for field conditions, Equation 8.3 may be used, but the Ce_t term must be corrected for the partial pressure of oxygen in the bubble. For instance, if pure oxygen is delivered at 1.0 atm pressure,

$$Ce'_t = Ce_t \times \frac{760 \text{ mm/Hg}}{159.2 \text{ mm/Hg}} \qquad 8.4$$

AERATION TANK

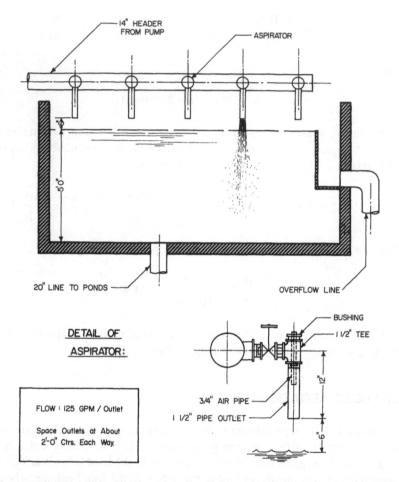

Figure 8.8. Burrows type aspirator system for aerating an entire hatchery water supply. (From Burrows, R. E. and Combs, B. D., *Progressive Fish-Culturist*, 30, 133, 1968. With permission.)

where Ce_t' = the solubility of oxygen with a pure oxygen atmosphere, Ce_t = the solubility of oxygen with an air atmosphere, 760 mm Hg = the partial pressure of oxygen in pure oxygen at 1.0 atm of pressure, and 159.2 mm Hg = the partial pressure of oxygen in air at a pressure of 1.0 atm. The corrected Ce_t' term for diffused air systems using compressed air is

$$Ce_t' = Ce_t \times \frac{\text{pressure of compressed air}}{\text{atmospheric pressure}} \qquad 8.5$$

The air pressure from the compressor may be given in pounds per square inch (psi). The conversion from psi to mm Hg is 0.0193 psi/mm Hg.

Oxygen can be purchased in bulk or in cylinders or can be produced on-site using a device called a pressure swing adsorption (PSA) unit. The PSA unit forces compressed air through a molecular sieve that adsorbs nitrogen, producing a gas that is 85 to 95% oxygen (Boyd and Watten 1989). A PSA system capable of generating 17 m^3/hr (25 kg/hr) costs $30,000 (Boerson and Chessney 1986). Operation of this unit would require a 40-HP compressor and backup electrical generator that would cost an additional $25,000. A smaller unit producing 2.4 m^3/hr (3 kg/hr) costs $5850 and the required 9.4-HP compressor could be purchased for $8100 (Watten 1991). Energy costs for the larger unit are calculated as follows, using an electrical power cost of $0.06/kWh. The cost of the gas produced is

$$40 \text{ HP} \times \frac{0.746 \text{ kW}}{\text{HP}} \times \frac{\$0.06}{\text{kWh}} = \frac{\$1.79}{\text{hr}} \times \frac{\text{hr}}{25 \text{ kg}} = \$0.07/\text{kg}$$

Similar calculations result in an energy cost of $0.14/kg of gas produced by the smaller unit.

Liquid oxygen is usually delivered by truck and stored in cryogenic vessels, which are leased or purchased. Storage tanks range in size from the 0.15-m^3 liquid capacity tanks commonly used on fish transport vehicles to 40 m^3. Each cubic meter of liquid oxygen provides 1145 kg of oxygen gas (Watten 1991). The cost of liquid oxygen is between $0.15 and $0.35/kg, depending on the trucking distance from the source (Watten 1991). A 5.68-m^3 storage tank can be leased for $250 to $350 per month or purchased for about $20,000 (Speece 1981).

Speece (1981) described five devices capable of transferring 90% or more of diffused oxygen to water under typical aquaculture conditions. They are enclosed packed column (Figure 8.9), U-tube (Figure 8.11), downflow bubble contact aerator (Figure 8.12), recycled diffused oxygenation (Figure 8.13), and rotating packed column (Figure 8.14) devices.

The dimensions and operation of a packed column (Figure 8.9) are given by Owsley (1981). A 25.4-cm diameter tube, 1.52 m tall, packed with 3.81-cm diameter plastic rings (Figure 8.10) is supplied with a water flow of 568 L/min. When oxygen was injected at the bottom of the column so that it rose counter-current to the influent water, the effluent DO concentration was in excess of 40 mg/L when influent DO was 6 mg/L and 90% oxygen absorption was achieved.

A U-tube (Figure 8.11) is a deep hole with a baffle in the center that forces water to the bottom, then back up to the surface. Gas is injected into the inlet side of the U-tube and the pressure-increase as the bubbles descend the tube enhances gas transfer. Speece (1981) reported that with a tube depth of 12.2 m, water velocity of 1.8 m/sec, and influent DO of 6 mg/L, effluent DO would be about 40 mg/L and 90% of the injected oxygen would be absorbed if a recycle hood

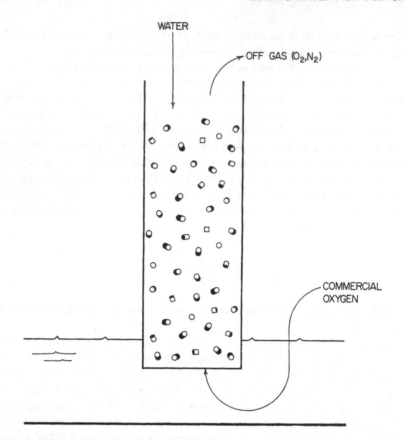

Figure 8.9. Enclosed packed column for absorption of oxygen. (From Speece, R. E., *Proceedings of the Bio-Engineering Symposium for Fish Culture,* Allen, L. J. and Kinney, E. C., Eds., American Fisheries Society, Bethesda, MD, 1981. With permission.)

is located at the effluent end to trap off gas and recycle it back down the tube. The head loss through the unit under these flow conditions would be 1.2 m.

The downflow bubble contact aerator (Figure 8.12) consists of a surface agitator at the top of a 3 m high cone. Oxygen is injected into the cone and the bubbles are trapped inside because the water velocity is greater at the top of the cone than at the bottom. Thus, the contact time of the bubbles in the water is greatly increased. Discharge DO from this device should be 22 mg/L when influent DO is 6 mg/L and 90% of the injected oxygen is absorbed (Speece 1981).

Speece (1981) also described two devices that can be placed directly into shallow raceways to achieve 90% oxygen adsorption. Recycled diffused oxygenation (Figure 8.13) achieves this by continuously recycling off gas back to the diffuser with a small blower. The rotating packed column (Figure 8.14) is a tube filled with plastic rings (Figure 8.10), half-submerged in the water column and rotating counter-current to the raceway flow. The unit is enclosed in a hood that contains a pure oxygen atmosphere.

Figure 8.10. Commercially available plastic rings used to facilitate gas transfer in packed columns.

Watten (1991) described a multi-stage column system in which the head requirement for oxygenation is reduced by employing a series of short columns. The column bank is operated in parallel with regard to water flow and in series relative to gas flow (Figure 8.15). Thus, off gas is repeatedly contacted with inlet water and high gas transfer is achieved through distances of water fall of 0.3 to 1.0 m. A patented device, called the Low Head Oxygenator (LHO), is available commercially and design details are proprietary.

Use of the devices described above result in effluent DO concentrations greatly in excess of saturation. Thus, only a sidestream of water needs to be oxygenated.

PROBLEMS WITH GAS SUPERSATURATION

Occasionally, groundwater supplies used for fish culture are supersaturated with nitrogen gas. This condition is often associated with undersaturation of oxygen. Supersaturation of dissolved nitrogen (DN) can only be accomplished by a change in pressure or temperature that lowers the solubility of the gasses in water. Groundwater becomes supersaturated with dissolved gasses when cool winter rains and snow melt enter a warmer aquifer. Dissolved oxygen levels may be reduced by microbial respiration in the soil. Water may be supersaturated with atmospheric gasses by a faulty pump that draws air in at the suction end or by heating water to accelerate fish growth. The changes in gas saturation associated with temperature changes may be expressed as

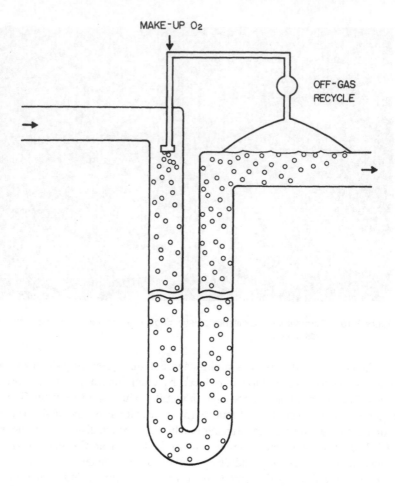

Figure 8.11. U-Tube with off-gas recycle for absorption of oxygen. [From Speece (1981).]

$$\text{Percent Saturation} = \frac{Ce_{t1}}{Ce_{t2}} \qquad 8.6$$

where Ce_{t1} = the gas concentration at temperature 1 and Ce_{t2} = the equilibrium concentration of that gas at temperature 2. Equilibrium concentrations of DN are presented in Table 8.2. Suppose water saturated with atmospheric gasses at 10°C is heated to 20°C. The resulting saturation of DO is 10.92/8.84 = 123%, and of DN is 18.14/14.88 = 122%.

When fish are exposed to supersaturated levels of DN they may experience a condition known as gas bubble disease in which nitrogen bubbles form in the blood and block capillaries by embolism. There is considerable evidence that only the inert gasses dissolved in water cause gas bubble disease. Exposure to supersaturated levels of DO or CO_2, which commonly occurs in fish transport

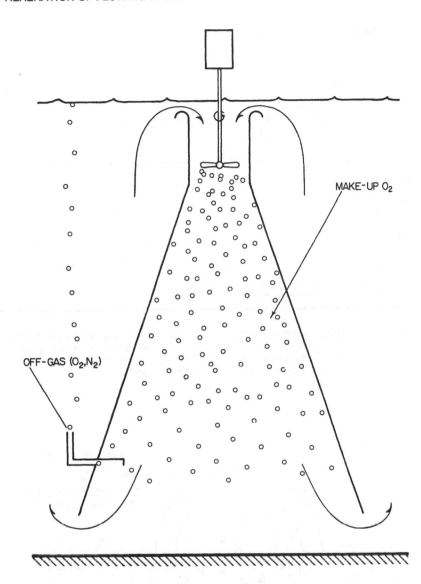

Figure 8.12. Downflow bubble contact aerator for absorption of oxygen. (From Speece, R. E., *Proceedings of the Bio-Engineering Symposium for Fish Culture*, Allen, L. J. and Kinney, E. C., Eds., American Fisheries Society, Bethesda, MD, 1981. With permission.)

and in static water aquacultures with high levels of photosynthesis and respiration, does not cause gas embolism. Apparently bubbles that form are metabolized before embolism occurs.

Bubbles can occur in fish blood only when the gas pressure in water exceeds the hydrostatic pressure. A 10-m submergence depth produces a hydraulic pressure of approximately 1 atm. For example, if the gas pressure in water is

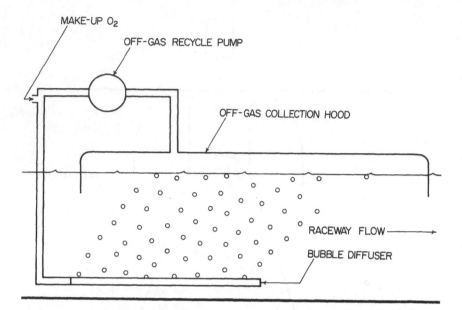

Figure 8.13. Recycled diffused oxygenation for absorption of oxygen in shallow ponds. (From Speece, R. E., *Proceedings of the Bio-Engineering Symposium for Fish Culture,* Allen, L. J. and Kinney, E. C., Eds., American Fisheries Society, Bethesda, MD, 1981. With permission.)

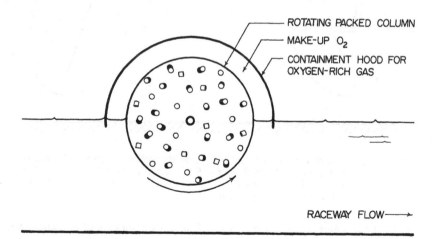

Figure 8.14. Rotating packed column for absorption of oxygen in shallow ponds. (From Speece, R. E., *Proceedings of the Bio-Engineering Symposium for Fish Culture,* Allen, L. J. and Kinney, E. C., Eds., American Fisheries Society, Bethesda, MD, 1981. With permission.)

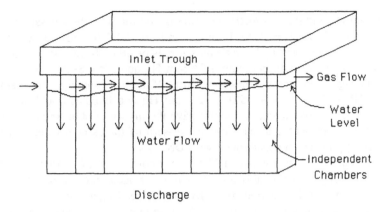

Discharge

Figure 8.15. Multi-stage column diffusion system in which water flows parallel and oxygen flows in series through a bank of short columns. (From Watten, B. J., *Engineering Aspects of Intensive Aquaculture,* Northeast Regional Agricultural Engineering Service, Ithaca, NY, 1991. With permission.)

Table 8.2   Equilibrium Concentrations in mg/L of Dissolved Nitrogen at T°C and 1.0 atm Pressure

T	0.0	0.1	0.2	0.3	0.4	0.5	0.6	0.7	0.8	0.9
4	20.82	20.77	20.72	20.67	20.62	20.57	20.52	20.47	20.42	20.38
5	20.33	20.28	20.23	20.18	20.13	20.09	20.04	19.99	19.85	19.90
6	19.85	19.81	19.76	19.71	19.67	19.62	19.56	19.53	19.49	19.44
7	19.40	19.35	19.31	19.26	19.22	19.18	19.13	19.09	19.05	19.00
8	18.96	18.92	18.88	18.83	18.79	18.75	18.71	18.67	18.63	18.58
9	18.54	18.50	18.46	18.42	18.38	18.34	18.30	18.26	18.22	18.18
10	18.14	18.10	18.06	18.02	17.99	17.95	17.91	17.87	17.83	17.79
11	17.76	17.72	17.68	17.64	17.61	17.57	17.53	17.50	17.46	17.42
12	17.39	17.35	17.31	17.28	17.24	17.21	17.17	17.13	17.10	17.06
13	17.03	16.99	16.96	16.93	16.89	16.86	16.82	16.79	16.75	16.72
14	16.69	16.65	16.62	16.59	16.55	16.52	16.49	16.45	16.42	16.39
15	16.36	16.32	16.29	16.26	16.23	16.20	16.16	16.13	16.10	16.07
16	16.04	16.01	15.98	15.95	15.91	15.88	15.85	15.82	15.79	15.76
17	15.73	15.70	15.67	15.64	15.61	15.58	15.55	15.52	15.50	15.47
18	15.44	15.41	15.38	15.35	15.32	15.29	15.27	15.24	15.21	15.18
19	15.15	15.12	15.10	15.07	15.04	15.01	14.99	14.96	14.93	14.91
20	14.88	14.85	14.82	14.80	14.77	14.74	14.72	14.69	14.67	14.64
21	14.61	14.59	14.56	14.53	14.51	14.48	14.46	14.43	14.41	14.38
22	14.36	14.33	14.31	14.28	14.26	14.23	14.21	14.18	14.16	14.13
23	14.11	14.08	14.06	14.04	14.01	13.99	13.96	13.94	13.92	13.89
24	13.87	13.85	13.82	13.80	13.78	13.75	13.73	13.71	13.68	13.66
25	13.64	13.61	13.59	13.57	13.55	13.52	13.50	13.48	13.46	13.44
26	13.41	13.39	13.37	13.35	13.33	13.30	13.28	13.26	13.24	13.22
27	13.20	13.17	13.15	13.13	13.11	13.09	13.07	13.05	13.03	13.01
28	12.99	12.96	12.94	12.92	12.90	12.88	12.86	12.84	12.82	12.80
29	12.78	12.76	12.74	12.72	12.70	12.68	12.66	12.64	12.62	12.60
30	12.58	12.50	12.54	12.53	12.51	12.49	12.47	12.45	12.43	12.41

From Colt, J. E., *Computation of Dissolved Gas Concentrations in Water as Functions of Temperature, Salinity, and Pressure,* American Fisheries Society Special Publication 14, Bethesda, MD, 1984. With permission.

890 mm Hg and the barometric pressure is 760 mm Hg, the gas saturation of the water is 890/760 = 117%. Bubbles can form if ΔP, the difference between total gas pressure and hydrostatic pressure, is positive. If the fish is at a depth of 1 m, the hydrostatic pressure is 76 mm Hg and ΔP = 890 − (760 + 76) = 54 mm Hg. Therefore, bubbles could form. If the fish submerges to a depth of 2 m, ΔP = 890 − (760 + 152) = −22 mm Hg and bubbles could not form. Unfortunately, fish in culture do not have the luxury of choosing a favorable depth and gas bubble disease can cause serious losses at hatcheries whose water supplies are supersaturated with DN.

The data on toxic levels of DN are very conflicting (Weitkamp and Katz 1980). Fish eggs are probably not affected by nitrogen supersaturation. Meekin and Turner (1974) reported mortality in steelhead eggs incubated in water having a total gas pressure of 112% of saturation, but factors other than supersaturation may have caused this mortality (Whitkamp and Katz 1980). In general, tolerance to supersaturation increases with fish age, with the most serious losses to fry and juveniles. Salmonids appear to be less tolerant of gas supersaturation than most other species. The critical level of DN saturation is generally considered to be 110% of saturation (Thurston et al. 1979).

Because supersaturation of water is an unstable condition that tends to return to equilibrium, aeration, or in this case de-aeration, is a logical remedy. Because of their relative solubilities, nitrogen is 1.51 times easier to transfer to water than is oxygen (Colt and Westers 1982). Thus, the mass transfer equation for DN removal may be written as

$$RT_{DN} = 1.51 \, RS_{DO} \frac{(C - Ce)\left(1.025^{T-20}\right)(0.85)}{14.88} \qquad 8.7$$

where RT_{DN} = DN removal at conditions at the aquaculture site, RS_{DO} = the oxygen transfer for a particular aerator at standard conditions, C = the DN concentration in mg/L at the aquaculture site, Ce = the equilibrium concentration of DN, and T = temperature in °C. The equilibrium concentration of DN at standard conditions is 14.88 mg/L (Table 8.2).

Suppose a mechanical aerator that transfers oxygen at 2 kg/kWh at standard conditions is used to remove excess DN from water whose temperature is 10°C, total gas pressure is 115% of saturation, and barometric pressure is 740 mm Hg. Then RS_{DO} = 2.0 kg/kWh, Ce = 18.14 × 740/760 = 17.66 mg/L, and C = 17.66 × 115% = 20.31 mg/L. Thus,

$$RT_{DN} = (1.51)(2.0)\frac{(20.31 - 17.66)(1.025^{-10})(0.85)}{14.88} = 0.36 \text{ kg DN/kWh}$$

Note that the actual nitrogen transfer from water to air is very low because the driving force (C − Ce) is small. Now suppose that a flow of 500 gpm is to be degassed with a desired effluent DN level of 110% saturation. The amount of DN that must be removed is 20.31 − 19.43 = 0.88 mg/L. Then

$$\frac{0.88 \text{ mg}}{\text{L}} \times \frac{3.78 \text{ L}}{\text{gal}} \times \frac{500 \text{ gal}}{\text{min}} \times \frac{\text{kg}}{10^6 \text{ mg}}$$
$$\times \frac{\text{kWh}}{0.36 \text{ kg}} \times \frac{60 \text{ min}}{\text{hr}} \times \frac{\text{HP}}{0.746 \text{ kWh}} = 0.37 \text{ HP}$$

is required.

Recently packed columns have been used successfully to remove excess DN from hatchery water supplies. Owsley (1979) reported that at 25.4-cm tube, 1.52 m high, filled to 1.37 m with 3.81-cm plastic rings, supplied with 379 to 568 L/min of water reduced DN levels from 130% to near 100% of saturation.

Speece (1981) reported that any device that successfully adds pure oxygen to water will result in DN levels of less than 100% of saturation. Thus, oxygenation is probably the best alternative for groundwaters that are supersaturated with DN and undersaturated with DO.

SAMPLE PROBLEMS

1. A hatchery with 53°F water at an elevation of 1900 feet above sea level has ponds arranged in series of three with a 12-in. drop between ponds. The series receives 450 gpm. The hatchery manager is requested to produce 10-in. brown trout. How many fish should be stocked in each pond of the series if a weir separates them? If lattice gravity aerators are installed, how much could the stocking rates be increased? If the fish are valued at $1.50/lb, what was the economic benefit of installing the lattices? Ignore the possible effects of ammonia.

2. At the same hatchery as Problem 1 the first pond in a series contains 20,000 6-in. trout. How many 3-in. trout to be grown to 6-in. should be stocked in the second pond in the series if an inclined corrugated sheet aerator is installed between the ponds?

3. Water at 11 cfs to be used as a hatchery water supply emerges from a spring containing zero dissolved oxygen. The hatchery site is 1/4 mile from the spring head and 20 ft lower in elevation. Devise a means of getting the DO to 90% of saturation before the water reaches the hatchery site.

4. At the same hatchery as in Problem 3, the water is used until the DO is reduced to 5 mg/L but the EPA will not allow the hatchery to discharge this water unless it contains at least 7 mg/L. There is no elevation difference between the last pond and the receiving water so a mechanical aerator must be installed. How large a unit is necessary if the actual oxygen transfer (when initial DO is 5 mg/L) is 0.6 lb/HPh? If the unit is electric and power costs $0.12/kWh what would be the annual operation cost of the aerator?

5. A 2-HP mechanical aerator is rated by the manufacturer to deliver oxygen at 2.5 lb/HPh. If this unit is placed in a raceway receiving 3000 gpm of 50°F water at a point where fish respiration has decreased the DO to 5 mg/L, how

much will the DO be increased below the unit? If a second unit is installed below the first, what will the DO be below it? Remember that the published oxygen transfer was determined under standard conditions.

6. A trout farmer has 1000 gpm of 58°F water at an elevation of 1500 feet above sea level. His facility contains two-step raceways but presently he does not raise any fish in the lower unit because the upstream ponds are stocked such that the effluent DO is 5 mg/L. There is no elevation difference between the first and second ponds. Determine the cost-benefit of mechanical reaeration to 90% of saturation so that fish can be produced in the lower ponds. Assume the units cost $12,000 and are depreciated over 10 years. Power costs $0.11/kWh and the profit margin on 11-in. fish is $0.25/lb. Is it economical to aerate in this case? Use 1 lb/HPh for actual transfer rate.

7. Design a Burrows type aspirated air system capable of reaerating a 5000-gpm water supply.

8. A 1000-gpm hatchery water supply containing 5 mg/L DO at 12°C is to be reaerated with compressed air using a 2-HP compressor. The oxygen transfer of the diffuser device, under standard conditions is 1.5 kg/kWh, and the air is delivered at a pressure of 3 atm. Calculate the effluent DO concentration. Is the compressor properly sized for the job? If not, recommend what size should be installed.

9. How much liquid oxygen would be required to reaerate the water supply described in Problem 8? Assume a 90% transfer of oxygen to water. Which mode of reaeration would be most economical for this example?

10. A well delivers 500 gpm of 9°C water at an elevation of 1000 feet above sea level. The DO concentration is 25% of saturation and the DN level 125% of saturation. What size mechanical aerator is required to achieve acceptable dissolved gas levels for fish culture?

REFERENCES

Boerson, G. and J. Chessney. 1986. *Engineering Considerations in Supplemental Oxygen*. Presented at Northwest Fish Culture Conference, Springfield, OR.

Boyd, C. E. and B. J. Watten. 1989. Aeration systems in aquaculture. *CRC Critical Reviews in Aquatic Sciences* 1: 425-472.

Burrows, R. E. and B. D. Combs. 1968. Controlled environments for salmon propagation. *Progressive Fish-Culturist* 30:123-136.

Chesness, J. L. and J. L. Stephens. 1971. A model study of gravity flow cascade aerators for catfish raceway systems. *Transactions of the American Society of Agricultural Engineers*. 14:1167-1169, 1174.

Colt. J. E. 1984. *Computation of Dissolved Gas Concentrations in Water as Functions of Temperature, Salinity, and Pressure*. American Fisheries Society Special Publication 14, Bethesda, MD.

Colt, J. E. and G. Tchobanoglous. 1979. Design of aeration systems for aquaculture. Pp 138-148 in L. J. Allen and E. C. Kinney (eds.) *Proceedings of the Bio-Engineering Symposium for Fish Culture*. Fish Culture Section Publication 1, American Fisheries Society, Bethesda, MD.

Colt, J. E. and H. Westers. 1982. Production of gas supersaturation by aeration. *Transactions of the American Fisheries Society* 111:342-360.

Downing, A. L. and G. A. Truesdale. 1955. Some factors affecting rate of solution of oxygen in water. *Journal of Applied Chemistry* 5: 570-581.

Eckenfelder, W. W., Jr. 1970. Oxygen transfer and aeration. Pp 1 - 12 in W. W. Eckenfelder (ed.) *Manual of Treatment Processes*. Vol. 1. Water Resources Management Series, Environmental Sciences Service Corporation, Stamford CT.

Haskell, D. C., R. O. Davies, and J. Reckahn. 1960. Factors in hatchery pond design. *New York Fish and Game Journal* 7: 113-129.

Mayo, R. D. 1979. A Technical and economic review of the use of reconditioned water in aquaculture. Pp 508-520 in T. V. R. Pillay and W. A. Dill (eds.) *Advances in Aquaculture*. FAO Technical Conference on Aquaculture, Kyoto, Japan.

Meekin, T. K. and B. K. Turner. 1974. Tolerance of salmonid eggs, juveniles and squawfish to supersaturated nitrogen. *Washington Department of Fisheries Technical Report* 12:78-126.

Owsley, D. E. 1981. Nitrogen gas removal using packed columns. Pp 71-82 in L. J. Allen and E. C. Kinney (eds.) *Proceedings of the Bio-Engineering Symposium for Fish Culture*. Fish Culture Section Publication 1, American Fisheries Society, Bethesda, MD.

Soderberg, R. W. 1982. Aeration of water supplies for fish culture in flowing water. *Progressive Fish-Culturist* 44: 89-93.

Soderberg, R. W., J. B. Flynn, and H. R. Schmittou. 1983. Effects of ammonia on growth and survival of rainbow trout in intensive static-water culture. *Transactions of the American Fisheries Society* 112:448-451.

Speece, R. E. 1973. Trout metabolism characteristics and the rational design of nitrification facilities for water reuse in hatcheries. *Transactions of the American Fisheries Society* 102:323-334.

Speece, R. E. 1981. Management of dissolved oxygen and nitrogen in fish hatchery waters. Pp 53-62 in L. J. Allen and E. C. Kinney (eds.) *Proceedings of the Bio-Engineering Symposium for Fish Culture*. Fish Culture Section Publication 1, American Fisheries Society, Bethesda, MD.

Tebbutt, T. H. Y. 1972. Some studies on reaeration in cascades. *Water Research* 6: 297-304.

Thurston, R. V., R. C. Russo, C. M. Fetterolf, Jr., T. A. Edsall, and Y. M. Barber, Jr. (eds.) 1979. *A Review of the EPA Red Book: Quality Criteria for Water*. Water Quality Section, American Fisheries Society, Bethesda, MD.

Watten, B. J. 1991. Application of pure oxygen in raceway culture systems. Pp 311-332 in *Engineering Aspects of Intensive Aquaculture*. Northeast Regional Agricultural Engineering Service, Cornell University, Ithaca, NY.

Weitkamp, D. E. and M. Katz. 1980. A review of dissolved gas supersaturation literature. *Transactions of the American Fisheries Society* 109: 659-702.

Westers, H. and K. M. Pratt. 1977. Rational design of hatcheries for intensive salmonid culture, based on metabolic characteristics. *Progressive Fish-Culturist* 39: 157-165.

Whipple, W., Jr., J. V. Hunter, B. Davidson, F. Dittman, and S. Yu. 1969. *Instream Aeration of Polluted Rivers*. Water Resources Institute, Rutgers University, New Brunswick, NJ.

CHAPTER 9

Ammonia Production and Toxicity

AMMONIA PRODUCTION RATE

Ammonia is the principle nitrogenous by-product of fish metabolism and is of importance in fish culture because it is toxic to fish in its un-ionized form. The origin of metabolic ammonia is the deamination of amino acids utilized as energy. A metabolic nitrogen budget allows for the estimation of the contribution of dietary protein to the accumulation of ammonia in the water. For 100 g of protein fed, approximately 40 g will be assimilated as fish flesh. If 20 g are undigested and 5 g are uneaten, 35 g will be metabolized as energy (Lovell 1989). Protein is approximately 16% nitrogen. Thus, fish consumption of 100 g of dietary protein will result in 35 g × 16%, or 5.6 g of ammonia being excreted. The ammonia production rate may be expressed as

$$A = 56P \qquad\qquad 9.1$$

where A = the ammonia production rate in grams of total ammonia nitrogen (TAN) per kilogram of food and P = the decimal fraction of protein in the diet. For example, feeding a 45% protein diet would result in the production of 25.2 g of ammonia-nitrogen per kilogram of food. In this book, concentrations of all nitrogenous species are expressed as mg/L of nitrogen, unless stated otherwise, which is consistent with most of the literature. Thus, 25.2 mg/L N × 18/14 = 32.4 mg/L NH_4^+. The formula weights of NH_4^+ and N are 18 and 14, respectively.

AMMONIA TOXICITY

Aqueous ammonia occurs in two molecular forms and the equilibrium between them is determined by pH, and to a lesser extent, temperature:

$$NH_3 \rightleftarrows NH_4^+$$

and

$$NH_3 - N + NH_4^+ - N = TAN$$

The un-ionized form, NH_3, is a gas and can freely pass the gill membrane. The rate and direction of passage depends upon the NH_3 concentration gradient between the fish's blood and the adjacent water. Un-ionized ammonia is toxic to fish while NH_4 is relatively nontoxic.

Chronic exposure to NH_4 damages fish gills, reducing the epithelial surface area available for gas exchange (Burrows 1964; Flis 1968; Bullock 1972; Larmoyeaux and Piper 1973; Smith and Piper 1975; Smart 1976; Burkhalter and Kaya 1977; Thurston et al. 1978; Soderberg et al. 1984a; Soderberg et al. 1984b; Soderberg 1985) (Figures 9.1 to 9.3). Snieszko and Hoffman (1963), Burrows (1964), Larmoyeaux and Piper (1973), Walters and Plumb (1980) and Soderberg et al. (1983) have reported that NH_3 exposure predisposes fish to disease. Reduction in fish growth caused by NH_3 exposure is widely documented (Brockway 1950; Kawamoto 1961; Burrows 1964; Smith and Piper 1975; Robinette 1976; Burkhalter and Kaya 1977; Colt and Tchobanoglous 1978; Soderberg et al. 1983). Gill damage caused by ammonia exposure may contribute to reduced growth by reducing oxygen consumption (Burrows 1964; Smith and Piper 1975) or the osmoregulatory cost of hyperventilation induced by reduced epithelial surface available for oxygen uptake (Lloyd and Orr 1969).

CALCULATION OF AMMONIA CONCENTRATION

The average daily TAN concentration in a fish-rearing unit is the total daily ammonia production in mg divided by the total daily water flow in L. For example, suppose that 1000 lb of fish are held in a pond receiving 100 gpm of water and are fed 1% of their body weight per day of 40%-protein food. The ammonia production rate is A = 56P = 56(.40) = 22.4 g/kg of food. The daily feed ration is 1000 lb × 1% = 10 lb = 4.55 kg, and the daily ammonia production is 22.4 g/kg × 4.55 kg = 101.92 g = 101,920 mg. The total daily water flow is 100 gal/min = 378 L/min × 1440 min/day = 544,320 L, and the average daily TAN concentration is 101,920 mg/544,320 L = 0.19 mg/L.

Analytical procedures do not differentiate between the two forms of ammonia in solution, and only one is of consequence to the fish culturist. Thus, it is important to be readily able to determine the fraction of NH_3 in solution at any temperature and pH. Then, if the level of NH_3 above which toxic effects occur is known, waters may be characterized according to their reuseability with respect to ammonia.

Emerson et al. (1975) present the following formula to calculate the acid dissociation constant, expressed as the negative log, for ammonia, based on the values of Bates and Pinching (1949):

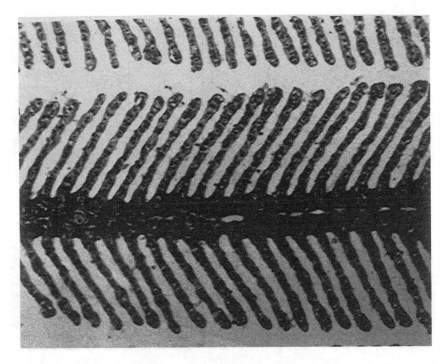

Figure 9.1. Normal rainbow trout gill magnified 75 ×. (From Soderberg, R. W., *Journal of Fish Diseases*, 8, 57, 1985. With permission.)

$$pKa = 0.09018 + \frac{2729.92}{T + 273.15} \qquad 9.2$$

where pKa = the negative log of the acid dissociation constant for ammonia and T = temperature in °C.

A formula to solve for the NH_3 fraction, $NH_3/(NH_3 + NH_4^+)$, in an ammonia solution is derived as follows. The acid dissociation for ammonia may be expressed as $NH_4^+ \rightleftarrows NH_3 + H^+$, and the equilibrium constant, Ka, is

$$Ka = \frac{(NH_3)(H^+)}{(NH_4^+)} = 10^{pKa}$$

Solving for the fraction $NH_3/(NH_3 + NH_4^+) = f$,

$$f = \frac{1}{10^{pKa-pH} + 1} \qquad 9.3$$

The un-ionized fraction, f, is the decimal fraction of NH_3 in an ammonia solution. Thus, $NH_3-N = TAN \times f$.

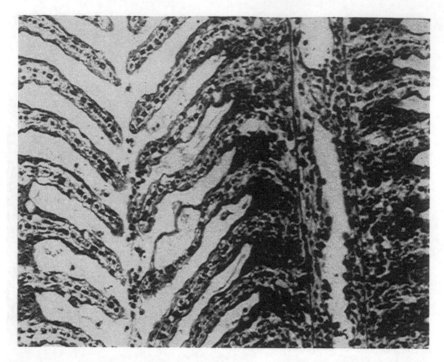

Figure 9.2. Rainbow trout gill with epithelial edema and hyperplasia at the bases of the filaments. These lesions are characteristic of ammonia exposure. Magnification is 150 ×. (From Soderberg, R. W., *Journal of Fish Diseases,* 8, 57, 1985. With permission.)

The use of these equations to calculate the un-ionized ammonia fraction is best illustrated by example. Suppose we wish to calculate f for 10°C water at pH values of 6, 7, and 8. From Equation 9.2,

$$pKa = 0.09018 + \frac{2729.92}{10 + 273.15} = 9.731$$

Substituting this into Equation 9.3 for a pH of 6,

$$f = \frac{1}{10^{9.731-6} + 1} = 0.000186$$

$$f \text{ at pH } 7 = \frac{1}{10^{9.731-7} + 1} = 0.00185$$

$$f \text{ at pH } 8 = \frac{1}{10^{9.731-8} + 1} = 0.0182$$

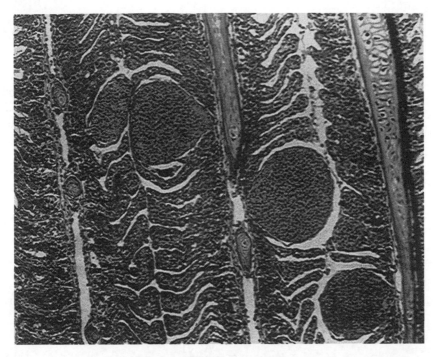

Figure 9.3. Rainbow trout gill with several large blood-filled aneurisms and hyperplasia causing fusion of some filiments. These lesions are characteristic of ammonia exposure. Magnification is 75×. (From Soderberg, R. W., *Journal of Fish Diseases,* 8, 57, 1985. With permission.)

Note that an increase in pH of one unit increases the un-ionized fraction of ammonia approximately 10-fold. Thus, for a given total ammonia level, fish in low-pH water will be exposed to less NH_3 than those in water with a higher pH. Because of its effect on the toxicity of ammonia, pH is one of the most important considerations in the selection of a water supply for intensive aquaculture.

Correction of the NH_3 Fraction for Ionic Strength

Ionic strength, as well as pH and temperature, control the ionization of ammonia and can significantly affect the un-ionized fraction of ammonia in brackish or marine waters. The apparent toxicity of NH_3 in saline waters is reduced because as ionic concentrations increase, electrostatic forces among ions and molecules increase the extent to which ionization occurs. Soderberg and Meade (1991) present the following procedure for correcting the NH_3 fraction in an ammonia solution for ionic strength. The strength of the electrostatic field that impinges on ionic activity is affected by the concentration and charge of ions in solution, and is expressed as the ionic strength (I):

$$I = \sum \frac{(M_i)(z_j)^2}{2} \qquad\qquad 9.4$$

where M is the molar concentration of a given ion (i) and z is its charge. For example, the ionic strength of a solution containing 100 mg/L each of Na^+ and Cl^- would be calculated as follows. Atomic weights are 23 for sodium and 35.5 for chlorine. Note that Na^+ and Cl^- are both monovalent (single charge) ions.

$$\frac{100 \text{ mg Na}}{L} \times \frac{\text{mole}}{23000 \text{ mg}} = 0.0043 \text{ M Na}$$

$$\frac{100 \text{ mg Cl}}{L} \times \frac{\text{mole}}{35500 \text{ mg}} = 0.0028 \text{ M Cl}$$

$$I = \frac{(0.0043)(1)^2}{2} + \frac{(0.0028)(1)^2}{2} = 0.0036 \text{ M}$$

To calculate the un-ionized fraction of ammonia in the above solution, the salinity correction term(s) is calculated from

$$s = -\frac{A'\sqrt{I}}{1+\sqrt{I}} \qquad\qquad 9.5$$

Values for the coefficient A′ are provided in Table 9.1. For the present example, if the pH is 7.5 and the temperature is 10°C,

$$s = -\frac{0.498\sqrt{0.0036}}{1+\sqrt{0.0036}} = -0.028$$

which is inserted into Equation 9.3 as follows:

$$f = \frac{1}{10^{pKa-pH-s}+1} \qquad\qquad 9.6$$

and

$$f = \frac{1}{10^{9.731-7.5+0.028}+1} = 0.00547$$

If f is not corrected for ionic strength,

Table 9.1. Values for the Coefficient A′ Used to Correct Ammonia Ionization for Ionic Strength

T, °C	A′	T, °C	A′
0	0.492	18	0.506
1	0.492	19	0.506
2	0.493	20	0.507
3	0.494	21	0.508
4	0.495	22	0.509
5	0.495	23	0.510
6	0.496	24	0.511
7	0.497	25	0.512
8	0.497	26	0.513
9	0.498	27	0.514
10	0.498	28	0.515
11	0.500	29	0.515
12	0.501	30	0.516
13	0.501	31	0.517
14	0.502	32	0.518
15	0.503	33	0.519
16	0.504	34	0.520
17	0.505	35	0.521

From Soderberg, R. W. and Meade, J. W., *Progressive Fish-Culturist*, 53, 119, 1991.

$$f = \frac{1}{10^{9.731-7.5} + 1} = 0.00584$$

and the un-ionized fraction is overestimated by 6.3%.

ASSIGNMENT OF AMMONIA MAXIMA FOR HATCHERIES

The maximum level of NH_3 at which the toxic effects are tolerable may be selected as a hatchery design criterion in a manner similar to that used to assign DO minima (Chapter 6). Robinette (1976) reported reduced growth of catfish at NH_3 concentrations greater than 0.12 mg/L. Colt and Tchobanoglous (1978) found that catfish growth was reduced at all concentrations above 0.048 mg/L, and that the growth rate at 0.517 mg/L was half of that at 0.048 mg/L. Larmoyeaux and Piper (1973) reported reduced growth of rainbow trout at 0.0166 mg/L NH_3, but not at 0.0125 mg/L. The U.S. Environmental Protection Agency (Thurston et al. 1978) has established an acceptable level of NH_3 for aquatic organisms of 0.016 mg/L. This is probably a reasonable design criterion for intensive aquaculture although catfish, and possibly other species, are apparently less sensitive to NH_3 than are trout, and higher NH_3 maxima may be acceptable.

CARRYING CAPACITY WITH RESPECT TO AMMONIA

Incorporation of ammonia toxicity considerations into hatchery design involves the calculation of the number of permissible water uses before NH_3 accumulates to the designated maximum level. A water use is defined as the carrying capacity, with respect to DO, of reaerated water. Consider the following hatchery data: carrying capacity in reaerated water = 10 lb/gpm, F = 1.2% body weight/day, P = 42%, pH = 7.7, T = 12°C, and maximum = 0.016 mg/L. Calculate the carrying capacity with respect to ammonia, considering unlimited aeration capacity. First, use Equation 9.1 and the appropriate units conversions to calculate the TAN concentration after one water use:

$$TAN = \frac{10 \text{ lb} \times 0.012 \times 56 \, (0.42) \, g \times \dfrac{kg}{2.2 \text{ lb}} \times \dfrac{1000 \text{ mg}}{g}}{\dfrac{1 \text{ gal}}{\min} \times \dfrac{3.78 \text{ L}}{gal} \times \dfrac{1440 \min}{day}}$$

TAN = 0.24 mg/L

Then use Equation 9.2 to calculate the pKa and Equation 9.3 to calculate the un-ionized fraction:

$$pKa = 0.09018 + \frac{2729.92}{273.15 + 12} = 9.664$$

$$f = \frac{1}{10^{9.664-7.7} + 1} = 0.0108$$

The NH_3 concentration after one water use is 0.24 mg/L × 0.0108 = 0.00259 mg/L, and the number of permissible water uses is 0.016/0.00259 = 6.18. Thus, the carrying capacity with respect to ammonia exposure is 10 lb/gpm × 6.18 = 61.8 lb/gpm.

PRODUCTION CAPACITY ASSESSMENT

Recommended NH_3 maxima, reported as NH_3–N, for fish hatcheries range from 0.0125 mg/L (Larmoyeaux and Piper 1973) to 0.016 mg/L (Willingham et al. 1979). Meade (1985) reviewed the literature on ammonia toxicity to fish and reported that site-specific water quality characteristics may significantly affect NH_3 toxicity, and thus the maximum safe exposure level. Soderberg and Meade (1992) summarized the water quality influences on the toxicity of NH_3 to fish. If the metabolite effects on cultured fish cannot be explained by NH_3 alone, a bioassay technique such as the Production Capacity Assessment (PCA) of Meade (1988) may be necessary for predicting hatchery carrying capacity.

The PCA procedure involves experimental rearing of fish in a series of five or more containers so that fish in each rearing unit are exposed to the accumulated metabolites from the upstream units. Because this is a procedure to estimate carrying capacity in terms of ammonia exposure, it is essential that the water be reaerated to at least 90% of saturation at the influent of each rearing container. The water flow through the series is adjusted so that the fish in each container remove 25 to 30% of the DO. Each container contains an equal weight of fish that are fed at the same rates. At the end of a 2 to 6 week period the oxygen consumption in each of the serial containers is determined by subtracting the effluent from the influent levels before feeding. Fish are weighed and the specific growth rate of fish in each container is regressed against the cumulative oxygen consumption. Specific growth rate is log total fish weight at the end of the growth trial minus log fish weight at the beginning of the growth trial, divided by the number of days in the growth trial multiplied by 100. The cumulative oxygen consumption at which specific growth is reduced to a predetermined minimum, expressed as a percentage of the specific growth in the first container, represents carrying capacity in terms of oxygen use. Meade (1988) defines this level as the Effective Cumulative Oxygen Consumption (ECOC). The recommended minimum allowable specific growth reduction is 50% of the observed maximum during the PCA test and the cumulative oxygen consumption that results in this level of growth reduction is defined as the $ECOC_{50}$ (Meade 1988). Thus, carrying capacity is defined as the weight of fish per unit flow of water that extracts the $ECOC_{50}$ of DO from the water supply.

Following is an example of the calculations required for the PCA procedure. Calculation of $ECOC_{50}$, based on the data in Table 9.2, is done in three steps, the results of which are shown in Table 9.3. The determination of cumulative oxygen consumption through the series of five rearing units is shown in Columns 2 and 3 (Table 9.3). Specific growth rate of fish in each unit is shown in Column 4 (Table 9.3). The final step is the determination, by linear regression, of the cumulative oxygen consumption at which growth is 50% of maximum ($ECOC_{50}$ in mg DO/L), which is completed by regressing the data in Column 3 (Table 9.3) on that in Column 4 (Table 9.3). Results follow:

Correlation coefficient = –0.985
Intercept = 14.3
Slope = –19.0
Specific growth rate for 50% of maximum growth = 0.28
$ECOC_{50}$ = –19.0 (0.28) + 14.3
$ECOC_{50}$ = 9.0 mg/L.

Carrying capacity is estimated by regressing cumulative fish load (Column 1, Table 9.2) on cumulative oxygen consumption (Column 3, Table 9.3) and

Table 9.2. Fish Weights and Dissolved Oxygen Concentrations in Five Serial Reuse Units

Serial unit (water use)	Weight (g)			DO (mg/L)	
	Starting	Ending	Cumulative	In	Out
1	4495	5386	5386	10.5	7.8
2	4505	5271	10657	9.9	6.7
3	4475	4938	15595	9.0	6.2
4	4372	4648	20243	9.0	6.6
5	3861	3897	24140	8.8	6.6

Note: The five units are those used in the example of the calculations in the Production Capacity Assessment (Meade 1988) procedure.
The test was run for 15 days.

Table 9.3. Calculation of Effective Cumulative Oxygen Consumption below which 50% of Maximum Growth is Achieved ($ECOC_{50}$), Based on Data in Table 9.1

Serial unit	Oxygen consumption		Specific growth rate
	Per unit	Cumulative	
1	2.7	2.7	0.561
2	3.2	5.9	0.488
3	2.8	8.7	0.305
4	2.4	11.1	0.190
5	2.2	13.3	0.029

Note: Specific growth rate is log total fish weight at the end of the growth trial minus log fish weight at the beginning of the growth trial, divided by the number of days in the growth trial × 100.

using the regression equation to calculate the fish biomass whose oxygen consumption would equal 9.0 mg/L. The regression equation resulting from the data presented in this example is cumulative fish load = 1781 × cumulative oxygen consumption + 354. When the $ECOC_{50}$ of 9.0 is substituted for cumulative oxygen, the calculated cumulative fish load is 16,383 g. If the flow through the PCA rearing units is 3.0 L/min, the carrying capacity for the water supply would be 16.4 kg for 3 L/min or 5.5 kg per L/min.

Soderberg and Meade (Unpublished) predicted carrying capacities of three hatchery water supplies, using the procedures described here for estimating the biomass of 10-g lake trout that would result in a maximum NH_3 exposure of 0.016 mg/L. Then they determined the carrying capacities of these water supplies by conducting PCA bioassays using 10-g lake trout. The objective of the experiment was to compare the results from the two available methods of calculating hatchery production capacity.

The water supplies tested were at Wellsboro (pH 7.0) and Lamar (pH 7.3), Pennsylvania and Leetown, West Virginia (pH 7.7). Ammonia production per unit of the 42% protein GR6-30 diet of the U.S. Fish and Wildlife Service was estimated from the dietary nitrogen budget of Lovell (1989). The ammonia production rate per kilogram of fish was based on the feeding rate of Haskell (1959) and a growth rate of 5.8 temperature units per centimeter of growth (Dwyer et al. 1981). The NH_3 fraction was calculated from Equation 9.3. Carrying capacity was calculated as the kilograms of fish per L/min of flow that would result in an effluent NH_3–N concentration of 0.016 mg/L.

Simultaneous PCA bioassays were conducted in triplicate at the three hatchery sites using the procedure of Meade (1988) with lake trout from a single lot of fish. Each of five bioassay containers per PCA replicate (15 containers total) at each location was stocked with 3600 g of fish, and each five-unit series was supplied with 3 L/min of water. Aeration was applied at the heads of each rearing container so that influent DO concentrations were at least 90% of saturation. The bioassays were conducted for 4 to 8 weeks at each of the three sites. Fish were fed the GR6-30 diet at rates determined from Haskell's (1959) feeding equation and a growth rate of 5.8 temperature units/cm (Dwyer et al. 1981).

At the end of each test, fish were weighed and the specific growth rates were calculated for each bioassay level. Specific growth rate was regressed against cumulative oxygen consumption and the $ECOC_{50}$ for each location was determined from the respective regression equations. Carrying capacities were calculated by computing the fish biomass whose oxygen consumption would equal the $ECOC_{50}$ value per L/min of water flow. For example, if a total fish load of 15 kg corresponded to an $ECOC_{50}$ of 9.0 mg/L DO in a flow of 3.0 L/min, the carrying capacity from the PCA procedure would be 5.0 kg per L/min.

The carrying capacity estimated from the mean $ECOC_{50}$ value was compared to that determined by the estimated NH_3 accumulation for each of the three water supplies using Chi-square analysis at the 95% confidence level.

Calculated carrying capacities, in kg per L/min, based on estimated NH_3 accumulation to 0.016 mg/L were 13.7 at Wellsboro, 8.7 at Lamar, and 3.5 at Leetown (Table 9.4). Carrying capacities based on the PCA bioassay determinations were 6.2 at Wellsboro, 7.5 at Lamar, and 3.0 at Leetown (Table 9.4). Calculated carrying capacities were nominally higher than those determined experimentally, but statistically different ($P < 0.05$) only for Wellsboro (Table 9.4). Calculated results consider only NH_3 toxicity, whereas the PCA method accounts for toxicities of other metabolites, the effects of chronic disease epizootics, water quality factors influencing NH_3 toxicity, and other unknown factors. Thus, the expected PCA values would be lower than those calculated from estimated NH_3 exposure. The fact that the actual carrying capacity at Wellsboro was significantly different from that predicted from NH_3 calculations indicates that, for some water supplies, factors other than NH_3 exposure constitute important limitations to fish production.

Table 9.4. Carrying Capacities for Lake Trout in Three Water Supplies Determined by Two Different Methods

Hatchery site	ECOC$_{50}$ (mg/L oxygen)	Carrying capacity (kg per L/min)	
		PCA	NH$_3$ Accumulation
Wellsboro	10.2	6.2	13.7
Lamar	10.9	7.6	8.7
Leetown	5.8	3.0	3.5

Note: The two methods gave statistically different results at Wellsboro, but not at the other two sites ($p < 0.05$). The ECOC$_{50}$ is the effective cumulative oxygen consumption below which fish fail to achieve at least 50% of the maximum observed growth.

SAMPLE PROBLEMS

1. A one-acre pond 5 ft deep contains 4000 lb of tilapia being fed a 25% protein diet at 3% of their body weight per day. The concentration of TAN in the morning, before feeding, is 1.0 mg/L. What will be the TAN concentration 24 hr later?

2. The pH of a catfish pond changes during the day because of the photosynthetic removal of CO_2 and its subsequent replacement by respiration. Assume that TAN is constant because the ammonia production rate equals the ammonia removal rate and that the water temperature is a constant 28°C. At dawn the pH is 6.5 and the NH$_3$ concentration is 0.002 mg/L. At dusk the pH has risen to 9.5. To what concentration of NH$_3$ are the fish now being exposed?

3. How many pounds of 12-in. trout can be reared in 100 gpm of 51°F water with a pH of 7.7? The maximum allowable NH$_3$ exposure is 0.013 mg/L, the protein content of the diet is 45% and unlimited aeration is available.

4. The Department of Natural Resources of a Great Lakes state has declared that water supplies for the production of 20-g chinook salmon smolts must allow for at least three complete water uses. Calculate the maximum permissible pH of water for salmon hatchery supplies for an area where the water is 49°F, the elevation is 600 feet above sea level, and the maximum allowable NH$_3$ level is 0.0125 mg/L. Salmon are fed a 48%-protein diet.

5. Consider a catfish farm using 500 gpm of geothermal water in a western state. The pH is 9.5, the temperature is 30°C, and the elevation is 5000 feet above sea level. Production raceways are built on a mountainside with gravity aeration between them. Assume that aeration releases all gaseous ammonia (NH$_3$) present in the water. If the first set of raceways is loaded to capacity with 500 g fish being fed a 32%-protein diet, what will be the concentration of NH$_3$ at the head of the second set of ponds?

6. A National Fish Hatchery in the West at an elevation of 2000 feet above sea level uses reservoir water with a pH of 6.8, heated to 58°F, to rear 6-in. steelhead smolts on a 48%-protein diet. The hatchery heats and discharges 5000 gpm and circulates 50,000 gpm through the rearing ponds and aerators. Thus, the water is used 10 times before it is discharged. Calculate the NH$_3$ concentration in the hatchery effluent. Does this facility require the installation of filters to remove ammonia to protect its fish from NH$_3$ toxicity?

7. A fish rearing facility on the American East Coast uses river water heated by an electric generating facility. The flow is 113 L/sec, the pH is 7.4, and the

temperature is 13°C. Liquid oxygen is injected along the raceway so that DO tensions are always over 150 mm Hg and rainbow trout are reared at a density of 176 kg/m^3. The raceway is 27 m long, 2.5 m wide, and 1 m deep. Calculate the NH$_3$ concentration at the tail of the raceway.

8. Trout are grown in a serial water reuse system in order to evaluate a water source for a hatchery. Each of five rearing units is stocked with 10 lb of 3-in. fish and the series receives 2 gpm of water. After 5 weeks influent and effluent DO concentrations of each rearing unit are measured and the fish are weighed. The following data are obtained:

Serial Unit	Fish Weight (lb)	Influent DO	Effluent DO
1	18.7	10	7.5
2	19.1	9.5	6.9
3	16.2	9.5	7.1
4	13.6	9.6	7.2
5	11.6	9.6	7.1

Calculate the ECOC$_{50}$ and the production potential of this water supply. When the hatchery is built, how many rearing units should be placed in series?

9. Calculate the production potential of the water supply tested in Problem 8 using the ammonia accumulation procedure. The hatchery will produce 3-in. trout for stocking. The temperature is 52°F and the pH is 7.5.

10. You are instructed to conduct a PCA bioassay to test the production potential of a water supply to be used for the intensive production of 14-in. catfish. Twenty liters per minute of 28°C water is available for the test. What size and how many fish should you stock in each PCA rearing unit?

REFERENCES

Bates, R. G. and G. D. Pinching. 1949. Acid dissociation constant of ammonia ion at 0 to 50° and the base strength of ammonia. *Journal of Research of the National Bureau of Standards* 42: 419-430.

Brockway, D. R. 1950. Metabolic products and their effects. *Progressive Fish-Culturist* 12: 127-129.

Bullock, G. L. 1972. *Studies on Selected Myxobacteria Pathogenic for Fishes and on Bacterial Gill Disease of Hatchery-Reared Salmonids.* U.S. Bureau of Sport Fisheries and Wildlife, Technical Paper 60.

Burkhalter, D. E. and C. M. Kaya. 1977. Effects of prolonged exposure to ammonia on fertilized eggs and sac fry of rainbow trout, *Salmo gairdneri. Transactions of the American Fisheries Society* 106: 470-475.

Burrows, R. E. 1964. *Effects of Accumulated Excretory Products on Hatchery-Reared Salmonids.* U.S. Fish and Wildlife Service Research Report 66.

Colt, J. and G. Tchobanoglous. 1978. Chronic exposure of channel catfish, *Ictalurus punctatus,* to ammonia: effects on growth and survival. *Aquaculture* 15: 353-372.

Dwyer, W. P., R. G. Piper, and C. E. Smith. 1981. *Lake Trout,* Salvelinus namaycush, *Growth Efficiency as Affected by Temperature.* U.S. Fish and Wildlife Service Fish Culture Development Station (Bozeman, Montana) Information Leaflet 22.

Emerson, K., R. C. Russo, R. E. Lund, and R. B. Thurston. 1975. Aqueous ammonia equilibrium calculations: effects of pH and temperature. *Journal of the Fisheries Research Board of Canada* 32: 2379-2383.

Flis, J. 1968. Anatomicohistopathological changes induced in carp, *Cyprinus carpio*, by ammonia water. II. Effects of subtoxic concentrations. *Acta Hydrobiologica* 10: 225-233.

Haskell, D. C. 1959. Trout growth in hatcheries. *New York Fish and Game Journal* 6: 205-237.

Kawamoto, N. Y. 1961. The influence of excretory substances of fishes on their own growth. *Progressive Fish-Culturist* 23: 26-29.

Larmoyeaux, J. C. and R. G. Piper. 1973. Effects of water reuse on rainbow trout in hatcheries. *Progressive Fish-Culturist* 35: 2-8.

Lloyd, R. and L. D. Orr. 1969. The diuretic response by rainbow trout to sub-lethal concentrations of ammonia. *Water Research* 3: 335-344.

Lovell, T. 1989. *Nutrition and Feeding of Fish*. Van Nostrand, New York, NY.

Meade, J. W. 1985. Allowable ammonia for fish culture. *Progressive Fish-Culturist* 47: 135-145.

Meade, J. W. 1988. A bioassay for production capacity assessment. *Aquacultural Engineering* 7: 139-146.

Robinette, H. R. 1976. Effects of selected sublethal levels of ammonia on the growth of channel catfish, *Ictalurus punctatus*. *Progressive Fish-Culturist* 38: 26-29.

Smart, G. 1976. The effects of ammonia exposure on the gill structure of rainbow trout, *Salmo gairdneri*. *Journal of Fish Biology* 8: 471-478.

Smith, C. E. and R. G. Piper. 1975. Lesions associated with the chronic exposure to ammonia. Pp 497-514 in W. E. Ribelin and G. Migaki (eds.) *The Pathology of Fishes*. University of Wisconsin Press, Madison, WI.

Snieszko, S. F. and G. L. Hoffman. 1963. Control of fish diseases. *Laboratory Animal Care* 13: 197-206.

Soderberg, R. W., J. B. Flynn, and H. R. Schmittou. 1983. Effects of ammonia on the growth and survival of rainbow trout in intensive static-water culture. *Transactions of the American Fisheries Society* 112: 448-451.

Soderberg, R. W., M. V. McGee, and C. E. Boyd. 1984a. Histology of cultured channel catfish, *Ictalurus punctatus* (Rafinesque). *Journal of Fish Biology* 24: 683-690.

Soderberg, R. W., M. V. McGee, J. M. Grizzle, and C. E. Boyd. 1984b. Comparative histology of rainbow trout and channel catfish in intensive static-water aquaculture. *Progressive Fish-Culturist* 46: 195-199.

Soderberg, R. W. 1985. Histopathology of rainbow trout, *Salmo gairdneri* (Richardson), exposed to diurnally fluctuating un-ionized ammonia levels in static-water ponds. *Journal of Fish Diseases* 8: 57-64.

Soderberg, R. W. and J. W. Meade. 1991. The effects of ionic strength on un-ionized ammonia concentration. *Progressive Fish-Culturist* 53: 118-120.

Soderberg, R. W. and J. W. Meade. 1992. Effects of sodium and calcium on acute toxicity of un-ionized ammonia to Atlantic salmon and lake trout. *Journal of Applied Aquaculture* 1: 83-92.

Soderberg, R. W. and J. W. Meade. (Unpublished). Estimates of ammonia accumulation and a capacity bioassay as predictors of hatchery carrying capacity.

Thurston, R. V., R. C. Russo, and C. E. Smith. 1978. Acute toxicity of ammonia and nitrite to cutthroat trout fry. *Transactions of the American Fisheries Society* 197: 361-368.

Thurston, R. V., R. C. Russo, C. M. Fetterolf, Jr., T. A. Edsall, and Y. M. Barber, Jr. (eds.). 1979. *A Review of the EPA Red Book: Quality Criteria for Water.* Water Quality Section, American Fisheries Society, Bethesda, MD.

Walters, G. R. and J. A. Plumb. 1980. Environmental stress and bacterial infection in channel catfish, *Ictalurus punctatus* Rafinesque. *Journal of Fish Biology* 17: 177-185.

Willingham, W. T., J. E. Colt, J. A. Fava, B. T. Hillaby, C. L. Ho, M. Katz, R. C. Russo, D. L. Swanson, and R. V. Thurston. 1979. Ammonia. Pp 6–18. In *A Review of the EPA Red Book: Quality Criteria for Water,* R. V. Thurston, R. C. Russo, C. M. Fetterolf, Jr., T. A. Edsall, and Y. M. Barber, Jr. (eds.), Water Quality Section, American Fisheries Society, Bethesda, MD.

CHAPTER 10

Water Recirculation

The most intensive level of aquaculture is fish production in recirculating systems that recondition and recycle all, or nearly all, of the water through the fish-rearing units. Recirculating systems are one of two general types depending upon the amount of water that is recycled. Water reuse systems discharge approximately 10% of their water after each cycle. This is equivalent to using the water ten times in a raceway series. In closed systems, virtually all the water is recycled and the water requirement is only that necessary to replace losses due to evaporation, spillage, and filter backwash.

The technology of fish culture in recirculating systems involves meeting the following environmental requirements of the fish in a self-contained unit:

1. Maintenance of DO tensions suitable for efficient respiration
2. Maintenance of safe levels of un-ionized ammonia NH^3
3. Maintenance of water reasonably clear of solid waste

Maintenance of suitable DO tensions is addressed in Chapter 8. The present chapter describes the filtering of ammonia and solid waste from fish culture effluents. Ammonia can be removed from water by pH elevation followed by air-stripping, ion exchange resins or zeolites, or by biological filtration. Nearly all aquaculture applications involve the later process in which ammonia is biologically oxidized to nitrate. Solid waste is removed from culture water by gravitational settling or mechanical filtration.

BIOLOGICAL FILTRATION

Microbiology

Biological filtration depends upon nitrification, a microbiological process by which autotrophic bacteria oxidize ammonium (NH_4^+) to nitrite (NO_2^-) and then to nitrate (NO_3^-). Ammonium is relatively nontoxic to fish, but its removal from culture water reduces the levels of its equilibrium product, ammonia (NH_3). The intermediate product, NO_2^-, is quite toxic to fish and is an important concern in recirculating aquaculture. Nitrate is essentially nontoxic to fish and is allowed to accumulate in recirculating systems.

Several genera of bacteria are known to oxidize ammonium and NO_2^-, but the most important are *Nitrosomonas* that oxidizes NH_4^+ to nitrite and *Nitrobacter* for the subsequent oxidation to NO_3^-. The nitrification rate is influenced by pH, DO, bicarbonate and ammonia levels, and temperature.

The literature on pH optima for *Nitrosomonas* and *Nitrobacter* are conflicting (Hochheimer 1990), probably because these bacteria adapt to the pH conditions to which they are exposed. Biological filters can probably operate from as low as pH 5 to as high as pH 10 if the bacteria are allowed to acclimate to these conditions and are not subjected to rapid pH changes (Wheaton et al. 1991).

The stoichiometric requirement when 1 g of ammonia nitrogen is nitrified to NO_3 is 4.18 to 4.33 g of oxygen. Dissolved oxygen concentrations lower than 1 to 2 mg/L can limit the nitrification rate (Water Pollution Control Federation 1983).

Nitrification reduces pH of culture water by consuming bicarbonate and producing hydrogen ions. The stoichiometric requirement for bicarbonate is 4.17 to 7.14 g, as $CaCO_3$, per gram of nitrogen oxidized from NH_4^+ to NO_2^-. Gujer and Boller (1986) showed that an alkalinity of at least 1.5 meq/L (75 mg/L $CaCO_3$) was required to provide sufficient bicarbonate to maintain the maximum rate of nitrification. Closed aquaculture systems require the addition of a buffer, usually sodium bicarbonate, to maintain suitable pH and alkalinity levels (Hochheimer 1991).

Un-ionized ammonia is toxic to nitrification bacteria and may inhibit the nitrification process. *Nitrobacter* is inhibited by NH_3 levels as low as 0.1 to 1.0 mg/L, while *Nitrosomonas* is more tolerant (Anthonisen et al. 1976). Nitrous acid (HNO_2), which is in equilibrium with NO_2^- in aqueous solution, is toxic to nitrification bacteria at concentrations as low as 0.22 mg/L (Anthonisen et al. 1976), but the equilibrium constant between HNO_2 and NO_2^- is so high $(10^{-3.14})$ that virtually no HNO_2 is present at pH values greater than 5.0. Thus, nitrite toxicity is of little concern in recirculating aquaculture as a nitrification inhibitor. It is, however, an important water quality factor affecting fish health in these systems.

Nitrification is most efficient at temperatures from 30 to 35°C (Kawai et al. 1965), but nitrification occurs at temperatures from −5°C (Jones and Morita 1985) to 42°C (Laudelout and Van Tichelen 1960) and nitrification bacteria will adapt to a wide range of temperatures. Several studies have documented a linear relationship between nitrification rate and temperature (Knowles et al. 1965; Downing 1965; Downing and Knowles 1966; Haug and McCarty 1971; Wortman 1990). Nitrification rates at temperatures between 40 and 70°F subjected to linear regression analysis in 5°F increments result in a high correlation (r = 0.998) and a useful expression:

$$\text{Nitrification Rate} = 0.000006T - 0.0002 \qquad 10.1$$

where nitrification rate is in lb N/ft^2 of media surface per day and $T =$ temperature in °F.

Filter start-up is the most critical period in recirculating aquaculture because ammonia reaches high levels while the *Nitrosomonas* colony is being established. Similarly, NO_2^- reaches high levels following this ammonia peak until the *Nitrobacter* population expands to accommodate its oxidation. Because of this, biofilters are usually started with inorganic ammonia at up to 15 mg/L (Hochheimer 1991), and allowed to become established before fish are stocked. Alternatively, a filter may be started with media from an active biofilter.

Biofilter Configuration

The biological filter is a housing containing media upon which the nitrification bacteria grow and through which effluent water from the fish-rearing containers passes. Media may be stone, oyster shell, plastic rings or plates, plastic pellets or beads, or any other material that will support bacterial growth. Media types are characterized by their specific surface area (available growth surface per unit volume) and void fraction. An ideal media has substantial specific surface to minimize the space requirement of the filter and a large void fraction to provide high hydraulic loadings and prevent clogging. Specific surface areas and void fractions of some commonly used biofilter media are listed in Table 10.1.

Biofilters are one of three general types. The water may enter the top of the filter and trickle or flow down through the media (Figure 10.1), or water may enter the bottom and flood up through the media (Figure 10.2). A third general type of biofilter is the biodrum or biodisk, which is a round-bottomed trough in which the filter media rotates alternatively in and out of the water that flows through the trough. A biodrum contains a cylindrical basket filled with media or media-bound into a cylindrical configuration. A biodisk (Figure 10.3) contains disks arranged on a shaft which turns over the length of the trough.

Trickling filters may be aerated by placing vents near the bottom of the filter housing to allow air flow counter-current to the water flow. Submerged filters may have air or oxygen lines near their inlets which may also aid in keeping the media distributed and free from fouling. A submerged filter containing small, heavy media such as plastic beads or sand is called a fluidized bed when sufficient head pressure is applied to suspend the media in the water flow. The fluidized bed design solves the problem of low void fractions of small, high specific surface area media by maintaining greater water flow around the media particles than could be accomplished by gravity or low head flow.

Hydraulic loadings to biofilters must be great enough to keep the media wet, yet not so great as to scour the the biological film from the media. Minimum and maximum hydraulic loads for plastic biofilter media are specified by the

Table 10.1. Specific Surface Areas and Void Fractions of Some
Biofilter Media Types Commonly Used for Aquaculture

Media type	Specific surface (ft^2/ft^3)	Void (%)
1–3 in. Stone	19	46
3.5-in. Plastic Rings[1]	27–31	95
1.5-in. Plastic Rings[1]	40	96
Rigid Plastic Module[1]	27	97
3/8-in. Styrofoam beads[2]	123	35
5/16-in. Polyethylene pellets[3]	200	30

[1] Biofilter media commercially available from several manufacturers.
[2] Commercial Styrofoam packing material.
[3] Polyethylene molding stock.

manufacturers. Minimum recommended flows generally range from 29 to 55 cubic meters of water per day per square meter of filter cross-sectional surface area (0.55 to 1.05 gpm/ft²). Hydraulic maxima range from 72 to 350 m³/day/m² (1.37 to 6.65 gpm/ft²)(Wheaton et al. 1991).

REMOVAL OF SOLID WASTE

Solid waste generated by fish fed dry diets is characterized by high organic content, low density and a broad size spectrum (Chen and Malone 1991). Solid material must be removed from recirculating aquaculture systems because it clouds the water, clogs filters and may be harmful to fish. Furthermore, the breakdown of organic waste produces ammonia and consumes oxygen. Deep deposits of organic matter may decompose anaerobically releasing toxic gasses. Solids removal technologies developed for wastewater treatment may not be appropriate for recirculating systems where small inefficiencies accumulate. The mechanical filters required for sufficient solids removal to maintain fish health in and satisfactory operation of recirculating aquaculture systems are expensive, require high pressure, and may waste considerable water for their backwash cycles.

Sedimentation

Wastewaters from conventional salmonid hatcheries are commonly treated by sedimentation in linear clarifiers prior to discharge to their receiving waters. Generally, only the cleaning waste from hatcheries using raceways needs to be treated to meet discharge regulations. A linear clarifier is a large raceway with a weir at the effluent end to provide surface discharge. A linear clarifier 100 ft long with a detention time of 2 hr and a water velocity of 0.056 ft/sec will remove 85% of the suspended solids produced by intensively reared fish (Jensen 1972).

Clarification by sedimentation alone may be appropriate for water reuse systems that exchange a substantial amount of water with each recirculation cycle. Closed systems are usually housed indoors where linear clarifiers would

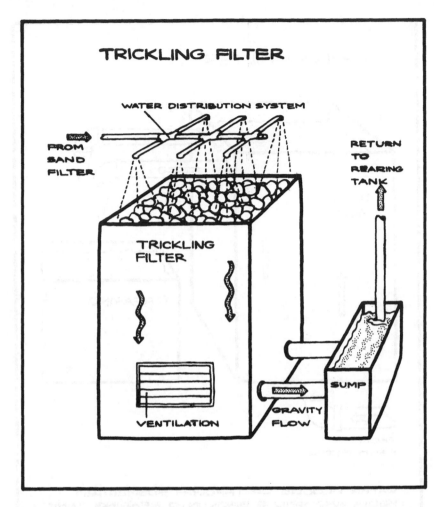

Figure 10.1. Diagram of trickling biological filter used for closed system aquaculture. (From Soderberg, R. W. and Quigley, J. T., *Technol. Perch Aquaculture*, University of Wisconsin Sea Grant Institute, 1977. With permission.)

require excessive space. Furthermore, sedimentation does not remove suspended particles smaller than 100 μm. Accumulated fine suspended particles in the flows of recirculating aquaculture systems may be directly toxic to fish (Chapman et al. 1987; Timmons et al. 1987).

The space requirement for sedimentation may be substantially reduced by use of a tube settler (EPA 1975; Soderberg and Quigley 1977) in which water is forced to pass upward through a bank of diagonal tubes (Figure 10.4). Soderberg and Quigley (1977) recommended a hydraulic capacity of 2 gpm/ft² of top surface of a commercially available tube settler module. In any

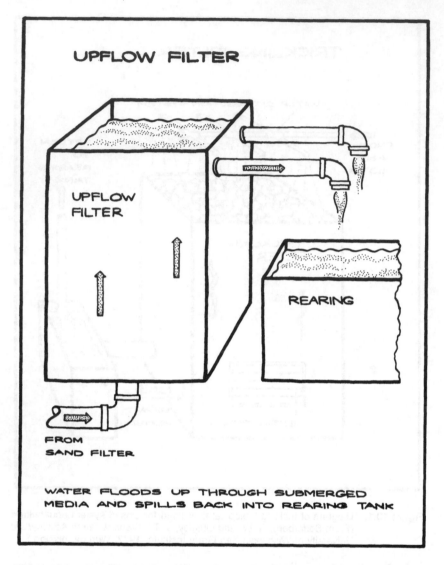

Figure 10.2. Diagram of submerged biological filter used for closed system aquaculture. (From Soderberg, R. W. and Quigley, J. T., *Technol. Perch Aquaculture*, University of Wisconsin Sea Grant Institute, 1977. With permission.)

sedimentation system, accumulated sludge must be regularly removed to prevent solids mineralization and anaerobic decomposition. Sedimentation with or without tube settlers is an insufficient means of solids removal for closed system aquaculture due to the failure to remove particles smaller than 100 μm.

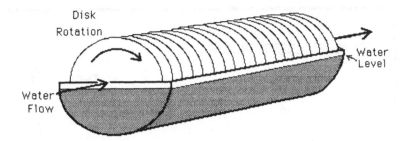

Figure 10.3. Diagram of biodisk biofilter used for closed system aquaculture.

Mechanical Filters

Rotating or inclined screens have been used to filter particles from fish culture effluents. The particle size removed is determined by the mesh size of the screen. Continuous cleaning is required to prevent clogging of screen filters.

Granular media filters such as swimming-pool sand filters (Figure 10.5) can effectively remove particles as small as 20 μm (The Task Committee on Design of Wastewater Filtration Facilities 1986). Sand filters have high head losses and therefore require large pumps to operate. Pump size requirements and flow rates through commercially available pool filters are listed in Table 10.2. Actual flow through these filters varies with backwash frequency and the backwash procedure consumes 75 to 100 gallons of water (Soderberg and Quigley 1977).

Diatomite earth (DE) filters and disposable cartridge filters can effectively remove particles as small as 1 μm (Chen and Malone 1991) and may be practical as polishing filters in closed systems if effective pre-filtration is accomplished.

Other possible mechanisms for fine solids removal in closed system aquaculture include foam fractionation in which suspended material is trapped on bubble surfaces and skimmed from the water as foam, and ozonation in which organic matter is oxidized by ozone gas, which can be generated on site.

Soderberg and Quigley (1977) and Chen and Malone (1991) concluded that a tube settler followed by a granular media filter was probably the most reasonable and cost-effective solids removal system for most closed aquaculture system applications (Figure 10.6). Problems remaining in the selection of this configuration include the space requirement, remineralization of solids from the accumulated sludge, head loss through the sand filter, and the failure to remove fine solids (Chen and Malone 1991).

Effective solids removal remains a major obstacle in the successful implementation of closed system aquaculture.

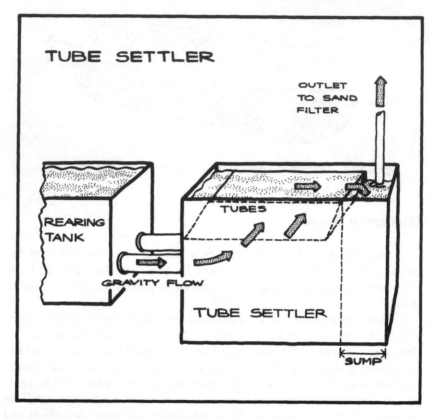

Figure 10.4. Diagram of tube settler used for closed system aquaculture. (From Soderberg, R. W. and Quigley, J. T., *Technol. Perch Aquaculture,* University of Wisconsin Sea Grant Institute, 1977. With permission.)

CONTROL OF NITRITE TOXICITY

Biofilter effluents contain residual levels of NO_2^- and nitrite exposure causes a disease in fish called methemoglobinemia in which functional anemia results from the conversion of hemoglobin to methemoglobin. Nitrite toxicity is related to the environmental chloride concentration, apparently because the gill epithelial chloride cells cannot distinguish between Cl^- and NO_2^-, and Cl^- inhibits NO_2^- absorption. Coho salmon exposed to 9 mg/L NO_2^- with a Cl^- concentration of 20 mg/L averaged a 64% conversion of their hemoglobin to methemoglobin, which resulted in 50% mortality. When fish were exposed to the same concentration of NO_2^- in a solution containing 148 mg/L Cl^-, none died and the incidence of methemoglobin conversion was 39%. Fish in this study survived an NO_2^- exposure of 30 mg/L when the Cl^- concentration was 261 mg/L (Perrone and Meade 1977). Thus, the amount of NO_2^- tolerated by fish is related to the Cl^- concentration, and fish in brackish or sea water are resistant to methemoglobinemia (Almendras 1987). Schwedler et al. (1985) measured the amount of methemoglobin

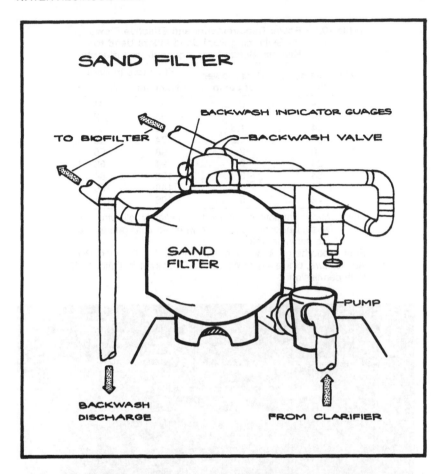

Figure 10.5. Diagram of swimming-pool sand filter used for closed system aquaculture. (From Soderberg, R. W. and Quigley, J. T., *Technol. Perch Aquaculture*, University of Wisconsin Sea Grant Institute, 1977. With permission.)

in channel catfish exposed to NO_2^- at NO_2^-:Cl^- ratios of 1:1 and 1:3. Fish in the low Cl^- concentration had 80% methemoglobin, but there was only 25% methemoglobin in the blood of the fish at the higher Cl^- concentration.

Nitrite toxicity is also related to calcium concentrations, probably because divalent cations decrease membrane permeability (Potts and Fleming 1974) and reduce NO_2^- absorption. Steelhead in solutions containing 67 mg/L Cl^- from NaCl exhibited a 96-hr LC_{50} of 0.97 mg/L of NO_2^-. When the same amount of Cl^- was added as $CaCl_2$, the 96-hr LC_{50} (median lethal concentration) of NO_2^- to steelhead was 22.7 mg/L (Wedemeyer and Yasutake 1978).

Sublethal levels of NO_2^- increase the susceptibility of fish to bacterial diseases (Hanson and Grizzle 1985) and NO_2^- exposures as low as 1.0 mg/L resulted in a 21% conversion of hemoglobin to methemoglobin in channel

Table 10.2. Power Requirements and Effective Flows
for Swimming Pool Sand Filters Used to
Remove Solids from Aquaculture Effluents

Filter diameter (in.)	Horsepower of pump	Flow rate in gpm	
		Maximum	Minimum
24	0.75	47	31
24	1	63	42
24	1.5	78	52
30	1.5	74	49
30	2	98	65
34	2	94	63
34	3	126	84
42	3	148	99
42	5	196	131

Note: Maximum flow is achieved immediately following backwash. Backwash is required when flow has decreased to
the minimum rate.

From Soderberg, R. W. and Quigley, J. T., *Technol. Perch
Aquaculture,* University of Wisconsin Sea Grant Institute, 1977.
With permission.

Figure 10.6.   A tube settler followed by swimming-pool sand filter (granular media filter)
used for solids removal in an experimental closed aquaculture system.

catfish (Tomasso et al. 1979). Because of the involvement of Cl⁻ and Ca^{2+} in
the toxicity of NO_2^- to fish, it is not possible to make recommendations on safe
levels of NO_2^- for aquaculture.

Chloride addition has been recommended to control NO_2^- toxicity in ponds.
Bowser et al. (1983) and Schwedler et al. (1985) recommended that the NO_2^-:Cl⁻
ratio be increased to 1:3 and Tucker et al. recommended a ratio of 1:6. All three

studies cited here reported nitrite on an NO_2^-, not an N, basis and the salt used was NaCl. Presumably, less Cl⁻ would be required to control methemoglobinemia if the calcium salt were used.

CLOSED SYSTEM DESIGN

The following example illustrates closed aquaculture system design. Suppose we wish to construct a closed system with a production capacity of 1000 lb of 10-in. (254-mm) trout at 15°C. The feeding rate is calculated from Equation 2.2:

$$F = \frac{3 \times C \times \Delta L \times 100}{L}$$

If the food conversion, C, is 1.7 and ΔL is calculated from the regression equation $\Delta L = -0.167 + 0.066$ T (Table 2.10),

$$F = \frac{3 \times 1.7 \times 0.823 \text{ mm} \times 100}{254 \text{ mm}} = 1.65\% \text{ body weight/day}$$

and the maximum daily feeding rate for the system would be $1000 \times 1.65\% =$ 16.5 lb of food. The water requirement is calculated from Equation 7.1:

$$\text{Loading Rate} = \frac{(Oa - Ob) \times 0.0545}{F}$$

When $Oa = 9.76$ mg/L (Equation 5.1 or Table 5.4) and $Ob = 4.60$ mg/L, which is 75 mm Hg, the maximum loading rate or carrying capacity is

$$\text{Carrying Capacity} - \frac{(9.76 - 4.60) \times 0.0545}{0.0165} = 17 \text{ lb/gpm}$$

and a flow of 59 gpm is required to meet the DO demands of 1000 lb of fish. The ammonia production rate is calculated from A = 56P (Equation 9.1) and if a 45%-protein diet is used, 1000 lb of fish will produce

$$\frac{56(0.45) \text{ g TAN}}{\text{kg feed}} \times \frac{\text{kg}}{2.2 \text{ lb}} \times \frac{16.5 \text{ lb feed}}{\text{day}} = \frac{189 \text{ g TAN}}{\text{day}} \times \frac{\text{lb}}{453.6 \text{ g}}$$

$$= 0.417 \text{ lb TAN/day}$$

Calculation of the nitrification rate from Equation 10.1 requires that the temperature be converted to the Fahrenheit scale and NR = 0.000006 (59) – 0.0002

$= 1.54 \times 10^{-4}$ lb TAN/ft^2 day. The surface area required to support a nitrification colony capable of oxidizing the ammonia produced by 1000 lb of fish is

$$\frac{0.417 \text{ lb TAN}}{\text{day}} \times \frac{\text{ft}^2 \text{ day}}{1.54 \times 10^{-4} \text{ lb}} = 2708 \text{ ft}^2$$

If a submerged biofilter is used with a plastic pellet media with a specific surface area of 200 ft^2/ft^3 (Table 10.1), 2708/200 = 13.5 ft^3 of biofilter media is required. If a tube settler and pool sand filter are selected for solids removal, the tube settler top surface area required is 59 gpm × 1 ft^2/2 gpm) = 29.5 ft^2 (Soderberg and Quigley 1977) and a 2-HP, 30-in. sand filter would be chosen (Table 10.2).

Completion of the system design would probably include a bulk liquid oxygen system to meet the oxygen demands of the fish and the biofilter, a NaHCO$_3$ metering system to maintain system pH and replace bicarbonate consumed by nitrification, possibly a chloride salt metering system to control nitrite toxicity, and probably a temperature control system. Daily maintenance would include feeding the fish, siphoning sludge from the rearing units and tube settler, backwashing the sand filter, and possibly cleaning the biofilter. Required daily water chemistry monitoring would include pH, total alkalinity, NH$_3$, NO$_2^-$, DO, and water temperature. Procedures for the analyses of water for these parameters are found in Boyd (1979).

SYSTEM APPLICATIONS AND CONFIGURATIONS

The best known application of recirculating aquaculture technology is the home aquarium in which conditions favorable for sustained fish health are maintained with an under-gravel filter operated by a diaphragm air pump. Air bubbles rise in the filter stand pipes providing an air lift that circulates the aquarium water through the gravel of the filter, nourishing a nitrification fauna. Solid waste is decomposed by organisms in the filter gravel. More sophisticated aquaria may contain DE or cartridge filters mounted externally to increase solids removal capacity.

Water reuse hatcheries have been built by state and federal fisheries agencies to conserve water that has been heated to accelerate fish growth. The most common application is for the production of smolts of anadromous salmonids in 1 year when ambient water temperatures would require 2 years of growth. These are usually outdoor facilities using raceways or rectangular circulating ponds (Figure 4.3). Solids are removed directly from the rearing units. Generally, the entire system flow is aerated through a Burrows type aspirator (Figure 8.2) or similar mechanism and passed through biofilters before returning to the rearing units. Ten percent of the system flow is continuously discharged and

replaced with fresh, heated water. Thus, 90% of the system flow is reconditioned water. It is important to note that the pH of the water supply may obviate
the need for ammonia removal. If the volume of fresh water added can sustain
the required fish production (equivalent to ten water uses in a 90% reuse
system) without exceeding designated NH_3 maxima, biofilters are not required.
Installation of biofilters in such a situation would be detrimental because of the
unnecessary production of NO_2^- and destruction of alkalinity, which would
require costly mineral additions to mitigate.

Closed aquaculture systems have been proposed for fish production in areas
of low water availability or unfavorable water temperatures for growth of the
desired fish species. A further justification for the development of closed
system technology has been to produce fish close to their markets to reduce
logistic problems associated with storage and transport of fish caught at sea or
produced in distant locations.

Closed system design eventually becomes a compromise between space
efficiency and energy requirements. In most proposed applications, these
factors are related. Suppose we wish to produce a warm-water food fish in a
large metropolitan area in the northeastern U.S. The environment is hostile to
our selected species so the water must be heated, and to conserve heat the
facility must be indoors and recirculating aquaculture technology must be
employed. The water pumping costs can be minimized by choosing low head
components for the production system. These may be circular fish-rearing units
that require less water flow than linear units, sedimentation to remove solids,
and a biodrum or biodisk for the biological filter. Water circulation in such a
system could be accomplished with a low-energy air lift pump. The disadvantage of such a design is that the system components require more space than
their higher head loss counterparts, which means a large building is required
and a considerable water surface is exposed, which increases heat loss and
evaporation. An alternative design might employ linear or vertical raceways
that require more water to be pumped in exchange for higher fish densities and
a substantially reduced floor space requirement. The space requirement for
solids removal could be reduced by using mechanical filters and biofiltration
could be accomplished with a high-pressure fluidized bed. In such a design the
fish production per unit of building space is high, but per unit of energy
consumption is low. In the low head case, energy consumption is minimized
at the expense of the space requirement. Actual system design would depend
upon the location and fish species chosen and probably would be a compromise
between the two extremes described above.

Because of the energy requirements and extreme complexity of water
quality management necessary to maintain sustained fish health, closed system
aquaculture will probably not be an economically feasible means of producing
food fish at most locations in the near future. The technology of water recon-
ditioning for aquaculture has changed little in the last 20 years and major

technological advances affecting the viability of water reuse are probably not forthcoming. The future of closed system aquaculture probably depends upon increasing costs associated with capture fisheries and depletion of high-value marine stocks, increasing costs and health concerns of fish transport, consumer demand for fish, and consumer concern with the quality and environmental contamination of fish produced in capture fisheries.

SAMPLE PROBLEMS

Solve Problems 1 to 7 concerning the closed system culture of 4000 kg of 600-g tilapia in water heated to 30°C.

1. What specific surface area of biofilter media is required? Design a biodisk filter for this system using 0.25-in. plastic sheeting to construct the disks.
2. Design an alternative trickling biofilter using 1.5-in. plastic rings with a hydraulic capacity of 2.5 gpm/ft^2.
3. Water entering the biofilter contains 75 mm Hg DO. Calculate the DO concentration of the biofilter effluent if no DO is added during passage through the biofilter. Is aeration of the biofilter required?
4. How much sodium bicarbonate (NaHCO$_3$) must be continuously metered into the system to replace alkalinity destroyed by nitrification?
5. Size a tube settler and sand filter for the system.
6. The residual nitrite level is 0.25 mg/L and fish are exhibiting clinical signs of methemoglobinemia. Calculate the concentration of NaCl that must be maintained in the system to control NO$_2^-$ toxicity.
7. Oxygen requirements for the system will be met with bulk liquid oxygen. How much must be delivered per month assuming an absorption efficiency of 90%?
8. Consider 90% water reuse systems for the production of 6-in. steelhead smolts. Water is heated to 58°F in order to complete the growth of these fish in 1 year. The hatcheries will be built in the American West where the elevation averages 1500 feet above sea level. Calculate the pH above which biological filters will be required.
9. A hatchery in the American Midwest circulates 2500 gpm of reused water, heated to 55°F to produce 40,000 lb/year of 6-in. lake trout. The water supply is a lake with a pH of 7.2 at an elevation of 600 feet above sea level. If no biofilters are used, how much new water must be continuously heated and added to the recirculating flow?
10. A closed system for the rearing of catfish has a carrying capacity of 10,000 kg of fish at 70°F. Calculate the carrying capacity if the water temperature is increased to 80°F. Assume that ample DO is available at both temperatures, there is sufficient rearing space for any additional fish, and the fish diets and feeding rates are the same for both temperatures.

REFERENCES

Almendras, J. M. E. 1987. Acute nitrite toxicity and methemoglobinemia in juvenile milkfish *Chanos chanos* Forsskal. *Aquaculture* 61: 33-40.

Anthonisen, A. C., R. C. Loehr, T. B. S Prakasam, and E. G. Srinath. 1976. Inhibition of nitrification by ammonia and nitrous acid. *Journal of the Water Pollution Control Federation* 48: 835-852.

Bowser, P. R., W. W. Falls, J. Van Zandt, N. Collier, and J. D. Phillips. 1983. Methemoglobinemia in channel catfish: Methods of prevention. *Progressive Fish-Culturist* 45: 154-158.

Boyd, C. E. 1979. *Water Quality in Warmwater Fish Ponds.* Alabama Agricultural Experiment Station, Auburn University, AL.

Chapman, P. E., J. D. Popham, J. Griffin, and J. Michaelson. 1987. Differentiation of physical from chemical toxicity in solid waste fish bioassay. *Water, Air, and Soil Pollution* 33: 295-308.

Chen, S. and R. F. Malone. 1991. Suspended solids control in recirculating aquacultural systems. Pp 170-186 in *Engineering Aspects of Intensive Aquaculture.* Northeast Regional Agricultural Engineering Service, Cornell University, Ithaca, NY.

Downing, A. L. 1965. *Advances in Water Quality Improvement,* Volume 1. University of Texas, Austin, TX.

Downing, A. L. and G. Knowles. 1966. Population dynamics in biological treatment plant. Third International Conference on Water Pollution Research. Munich, Germany.

Environmental Protection Agency (EPA). 1975. Process design manual for suspended solids removal. *U.S. EPA Technology Transfer Publication.*

Gujer, W. and M. Boller. 1986. Design of a nitrifying tertiary trickling filter based on theoretical concepts. *Water Research* 20: 1353-1362.

Hanson, L. A. and J. M. Grizzle. 1985. Nitrite-induced predisposition of channel catfish to bacterial diseases. *Progressive Fish-Culturist* 47: 98-101.

Haug, R. T. and P. L. McCarty. 1971. *Nitrification with the Submerged Filter.* Technical Report Number 149, Department of Civil Engineering, Stanford University, Stanford, CA.

Hochheimer, J. N. 1990. *Trickling Filter Model for Closed System Aquaculture.* Unpublished dissertation, University of Maryland, College Park, MD.

Hochheimer, J. N. 1991. Understanding biofilters, practical microbiology for ammonia removal in aquaculture. Pp 57-79 in *Engineering Aspects of Intensive Aquaculture.* Northeast Regional Agricultural Engineering Service, Cornell University, Ithaca, NY.

Jensen, R. 1972. Taking care of wastes from the trout farm. *American Fishes and U.S. Trout News* 16: 4-6, 21.

Jones, R. D. and R. Y. Morita. 1985. Low temperature growth and whole cell kinetics of a marine ammonium oxidizer. *Marine Ecology Progress Series* 21: 239-243.

Kawai, A., Y. Yoshida, and M. Kimata. 1965. Biochemical studies on the bacteria in the aquarium with a circulation system. II. Nitrifying activity of the filter sand. *Bulletin of the Japanese Society of Scientific Fisheries* 31: 65-71.

Knowles, G., A. C. Downing, and M. J. Barrett. 1965. Determination of kinetic constant for nitrifying bacteria in mixed culture with the aid of an electronic computer. *Journal of Genetic Microbiology* 38: 263-278

Laudelout. H. and L. Van Tichelen. 1960. Kinetics of the nitrite oxidation by *Nitrobacter winogradsky*. *Journal of Bacteriology* 79: 39-42.

Perrone, S. J. and T. L. Meade. 1977. Protective effect of chloride on nitrite toxicity to coho salmon, *Oncorhynchus kisutch*. *Journal of the Fisheries Research Board of Canada* 34: 486-492.

Potts, W. T. W. and W. R. Fleming. 1974. The effects of prolactin and divalent ions on the permeability to water of *Fundulus kansae*. *Journal of Experimental Biology* 53: 317-327.

Schwedler, T. E., C. S. Tucker, and M. H. Beleau. 1985. Non-infectious diseases. Pp 497-541 in C. S. Tucker (ed.) *Channel Catfish Culture*. Elsevier, New York, NY.

Soderberg, R. W. and J. T. Quigley. 1977. *The Technology of Perch Aquaculture*. University of Wisconsin Sea Grant College Program, Marine Advisory Report 416. Madison, WI.

The Task Committee on Design of Wastewater Filtration Facilities. 1986. Tertiary filtration of wastewaters. *Journal of Environmental Engineering* 112: 1008-1025.

Timmons, M. B., W. D. Youngs, J. M. Regenstein, G. A. German, P. R. Bowser, and C. A. Bisogni. 1987. *A Systems Approach to the Development of an Integrated Trout Industry for New York State*. New York State Department of Agriculture and Markets, Cornell University, Ithaca, NY.

Tomasso, J. R., B. A. Simco, and K. B. Davis. 1979. Chloride inhibition of nitrite induced methemoglobinemia in channel catfish *Ictalurus punctatus*. *Journal of the Fisheries Research Board of Canada* 36: 1141-1144.

Water Pollution Control Federation. 1983. *Nutrient Control, Manual of Practice*. Publication Number FD-7. Washington, D.C.

Wedemeyer, G. A. and W. T. Yasutake. 1978. Prevention and treatment of nitrite toxicity in juvenile steelhead trout, *Salmo gairdneri*. *Journal of the Fisheries Research Board of Canada* 35: 822-827.

Wheaton, F. W., J. N. Hochheimer, G. E. Kaiser, and M. J. Krones. 1991. Principles of biological filtration. Pp 1-31 in *Engineering Aspects of Intensive Aquaculture*. Northeast Regional Agricultural Engineering Service, Cornell University, Ithaca, NY.

Wortman, B. 1990. *Effect of Temperature on Biodrum Nitrification*. Unpublished Thesis. University of Maryland, College Park, MD.

Appendices

SOLUTIONS TO PROBLEMS

Chapter 2

1. MTU in 3 months = 12 + 13 + 11 = 36. Rainbow trout grow at 7.1 MTU/cm so in 3 months fish will grow 5.07 cm = 1.996 in. Fish are 9 in. = 0.2916 lb and will be 10.996 in. = 0.5318 lb. Anticipated gain per fish = 0.2402 lb × 20,000 fish = 4804 lb × 1.7 = 8167 lb feed = **164 bags of feed.**

2. 8.7 in. of growth are required. There are 12 MTU/month and 7.1 MTU are required per centimeter of growth. Required time = 13.07 months or 392 days + 14 days for swim up. Eggs should be delivered about **February 20.**

3. 0.446 lb of feed is required for 0.343 lb of gain. **Feed cost = 9.6 cents.**

4. ΔL = 0.0956 in. because growth in 90 days is 8.6 in. HC = 3 × 1.7 × 0.0956 × 100 = **48.7.**

5. 9 in. = 0.117 lb × 100,000 = 11,664 lb fish = **19,829 lb feed.**

6. C = 198,000/150,000 = 1.32. HC = 3 × 1.32 × ΔL × 100 = 8.55, so ΔL = 0.021 in. Lot 1: August 1, fish are 0.011 lb = 3.029 in. August 15, fish are 0.0149 lb, 3.344 in. 2-week gain = 336.5 lb. F = 8.55/3.344 = 2.56%. Feed requirement is **8.6 lb** on August 15. Lot 2: feed requirement is **25.99 lb.** Lot 3: feed requirement is **104.5 lb.**

7. a) HC = 3 × 1.5 × 0.0215 × 100 = **9.675.**
   b) F = 9.675/9.631 × 0.357 × 215,000 = **771 lb of feed.**
   c) F = 9.675/10.276 × 0.434 × 215,000 = **879 lb of feed.**

8. 0.341 lb gain per fish × 110,000 fish × 1.7 = **63,767 lb of feed.**

9. A: $\dfrac{\$370}{\text{ton feed}} \times \dfrac{1.85 \text{ ton feed}}{\text{ton gain}} = \dfrac{\$684.50}{\text{ton gain}}$.

   B: ton of gain costs $645.00.
   C: ton of gain costs $703.25.

10. A: 380 g protein/1000 × 3.9 C/g = 1482 C. 50 g fat/1000 × 8.0 C/g = 400 C. 160 g CHO/1000 × 1.6 C/g = 256 C. Total C in 1000 g of feed = 2138. C = 3850/2138 = 1.80 × $0.46/kg = $0.828/kg of gain.
    B: Cost is $0.829/kg of gain.
    C: Cost is $0.768/kg of gain.
    D: Cost is $0.794/kg of gain.

Chapter 3

1. $1000 \text{ acre} \times \dfrac{43,560 \text{ ft}^2}{\text{acre}} \times \dfrac{40 \text{ in.}}{\text{year}} \times \dfrac{\text{ft}}{12 \text{ in.}} \times \dfrac{7.48 \text{ gal}}{\text{ft}^3} \times \dfrac{\text{year}}{525,600 \text{ min}} \times 10\%$

    $= \textbf{207 gpm.}$

2. $\dfrac{5,000,000 \text{ ft}^3}{50,000 \text{ acre} \times \dfrac{43,560 \text{ ft}^2}{\text{acre}} \times 0.025 \text{ ft}} = \textbf{9.2\%.}$

3. $100 \text{ mile}^2 \times \dfrac{(5280)^2 \text{ ft}^2}{\text{mile}^2} \times \dfrac{1.667 \text{ ft}}{\text{year}} \times \dfrac{7.48 \text{ gal}}{\text{ft}^3} \times \dfrac{\text{year}}{525,000 \text{ min}} \times 65\%$

    $= \textbf{43,039 gpm.}$

4. $\dfrac{1000 \text{ gal}}{\text{min}} \times \dfrac{8.34 \text{ lb}}{\text{gal}} \times 120 \text{ ft} \times \dfrac{\text{HP min}}{33,000 \text{ ft lb}} \times \dfrac{0.746 \text{ kW}}{\text{HP}} \times \dfrac{8760 \text{ hr}}{\text{year}} \times \dfrac{\$0.065}{\text{kWh}}$

    $\div 0.8 = \textbf{\$16,102.74.}$

5. $100 \text{ HP} \times \dfrac{33,000 \text{ ft lb}}{\text{HP min}} \times \dfrac{\text{min}}{1200 \text{ gal}} \times \dfrac{\text{gal}}{8.34 \text{ lb}} = 330 \text{ ft.}$

    Draw down $= 330 - 120 = \textbf{210 ft.}$

6. $A = \dfrac{Q}{V}$ (Equation 3.1) $= \dfrac{0.453 \text{ m}^3/\text{sec}}{0.5 \text{ m}/\text{sec}} = 0.907 \text{ m}^2.$ The area of a trapezoid

    is $A = hb + h(c-b/2)$ where h, b, and c are the lengths of the height, base, and crown, respectively. For a trapezoid with a 2:1 side slope, $c-b = 4h$. If we let the base $= 2h$, the cross-sectional area to flow is $h(2h) + h(2h) = 0.907 \text{ m}^2$ and $h = 0.476$ m. **The dimensions of the channel are 0.476 m, 0.952 m and 2.856 m (h:b:c).**

7. Try 1-in. (1/12 ft) pipe:

    $K = \left[6.304 \left(2 \log \dfrac{1}{10^{-6}} + 1.14 \right) 1/12^{2.5} \right] = 0.166 \text{ cfs.}$

    $Q = 0.166 \sqrt{\dfrac{1.5}{200 + 99(1/12)}} = 0.014 \text{ cfs} = 6.3 \text{ gpm.}$

    Try 2-in. pipe: K = 0.939, Q = 35.1 gpm.
    Try 2.5-in. pipe: K = 1.64, Q = 60.7 gpm.
    Select the 2.5-in. pipe.

8. $V = \dfrac{1}{0.015} \left[\dfrac{(0.304)(0.610)}{2(0.304)+0.610} \right]^{2/3} \left(\dfrac{1}{1000} \right)^{1/3} = 1.9 \text{ m}/\text{sec.}$

    $Q = (1.9)(0.304)(0.610) = 0.35 \text{ m}^3/\text{sec} = \textbf{5580 gpm.}$

9. $n_s = \dfrac{1800\sqrt{50}}{175^{0.75}} = \textbf{264 rpm}$.

   Power requirement is 2 HP ÷ 0.8 = **2.5 HP.**
10. From 3.7, $Q = 1.37(0.5)(0.2)^{0.67} = \textbf{0.23 m}^3\textbf{/sec}$.

Chapter 4

1. $\dfrac{600\text{ gal}}{\text{min}} \times 15\text{ min}\dfrac{\text{ft}^3}{7.48\text{ gal}} = 1203\text{ ft}^3 = $ required volume.

   If $D = x$, $V = 1203 = (x)(3x)(30x) = 90x^3$.
   D = 2.37 ft, W = 7.12 ft, L = 71.2 ft.

2. $v = \dfrac{Q}{D \times W} = \dfrac{1.337\text{ ft}^3/\text{sec}}{(2.37)(7.12)\text{ ft}^2} = \textbf{0.079 ft/sec}$ (too slow).

   Set $v = 0.1$ ft/sec and calculate new depth; $0.1 = \dfrac{1.337}{D \times 7.12}$ and D = 1.88 ft,
   but there are not 4 exchanges per hour. Increase length so that $r = 4$; 1203 =
   L × 7.12 × 1.88 and L = 89.9 ft and v still = 0.1 ft/sec. If V = 0.1 and
   dimensions remain 30:3:1, the raceway volume is 953 ft³ and R = 5. All
   designs are satisfactory.
3. Flow is **718 gpm** and **v = 0.067 ft/sec** with 6 min/exchange. To increase
   velocity to 0.1 ft/sec, **decrease depth to 2 ft.** The new exchange rate is 4 min/
   exchange **(R = 15).**
4. Exchange rate = 3.58 min/exchange **(R = 16.7).**

   $v = Q/\pi r^2 = \dfrac{0.027\text{ ft}^3/\text{sec}}{2.63\text{ ft}^2} = \textbf{0.010 ft/sec.}$

5. $V = \pi r^2 h$. Q = 785 L/min **(208 gpm).**
6. If the depth is set at 0.6 m, the required radius is 5.35 m when R = 2.
7. The floor requires 17.98 m³ and if the pond is built with 30 cm of freeboard,
   the walls require 6.05 m³. 24.03 m³ = **$2403.**
8. A raceway that receives 30 L/sec of water, has a velocity of 0.033 m/sec and
   R = 4 has the dimensions of 30.00:2.02:0.45 m (L:W:D). Allowing for 30 cm
   of freeboard, the concrete requirement is 13.14 m³ = **$1314.**
9. Circular pond, **39 kg/m³**; raceway, **78 kg/m³.**
10. Exchange rate = 37 min/exchange **(R = 1.62)**

Chapter 5

1. $22.4 \times \dfrac{293.15}{273.15} = 24.0\text{ mol}/\text{L}.$

   $\dfrac{946{,}000}{1.07 \times 10^6} \times \dfrac{\text{mol}}{24.0\text{ L}} \times 0.00033 = \textbf{0.53 mg/L.}$

2. a) $Ce = 14.161 - 0.3943(20) + 0.0077147(20^2) - 0.0000646(20^3)$

$\times \dfrac{760}{760 + \dfrac{1500}{32.8}} = \textbf{8.34 mg/L.}$

b) $Ce = 7.33 \times \dfrac{760}{760 + \dfrac{-700}{32.8}} = \textbf{7.54 mg/L.}$

c) $Ce = 10.92 \times \dfrac{29.90}{29.92} = \textbf{10.91 mg/L.}$

d) $Ce = \textbf{8.84 mg/L.}$

3. a) $\dfrac{5}{10.56} = \textbf{47.4\% Saturation} \times 734.6 \times 0.20946 = \textbf{72.9 mm Hg.}$

b) $\dfrac{12}{7.59 \times 0.976} = \textbf{161.9\% Saturation} \times 155.4 = \textbf{251.7 mm Hg.}$

c) $\dfrac{10.92 \times 2.3}{10.30} = \textbf{243.8\% Saturation} \times 150.1 = \textbf{366.1 mm Hg.}$

d) $\dfrac{10.56}{8.55} = \textbf{123.5\% Saturation} \times 153.9 = \textbf{190.1 mm Hg.}$

4. $\dfrac{760}{2} = \dfrac{760^2}{760 + \dfrac{E}{32.8}}$. $E = \textbf{24,928 feet above sea level.}$

5. $Ce = 7.75 - \left[17(0.0841) - 0.00256(28) + 0.0000374(28^2) \right] = \textbf{7.04 mg/L.}$

6. $\dfrac{7.87 \text{ mg}}{L} \times \dfrac{28.31 \text{ L}}{ft^3} \times \dfrac{(5280^3) \text{ ft}^3}{mile^3} \times \dfrac{kg}{10^6 \text{ mg}} = \textbf{32,795,660 kg.}$

7. $8.11 \times \dfrac{760}{159.2} = \textbf{38.72 mg/L.}$

8. BP = 713.09 mm Hg.
    VP at 10% RH = 0.92 mm Hg and at 90% RH VP = 8.38 mm Hg.
    $PO_2 w = \dfrac{5}{10.29} \times 149.17 = 72.77$ mm Hg at 10% RH

    $PO_2 w = \dfrac{5}{10.29} \times 147.61 = 71.72$ mm Hg at 90% RH

    Difference = 1.05 mm Hg.

9. a) Percent error $= 1 - \dfrac{0.20946 \,(726.54 - 4.6)}{0.20946 \,(726.54)} = \textbf{0.6\%.}$

b) Percent error $= 1 - \dfrac{\dfrac{5}{10.30} \times 0.20946\,(716.86 - 35.7)}{\dfrac{5}{10.30} \times 0.20946\,(716.86)} = \mathbf{5.0\%.}$

c) Percent error = **4.2%.**
d) Percent error = **5.8%.**

10. $\dfrac{3.5}{7.75} \times 159.2 = 159.2 \times \dfrac{DO}{11.47} DO = \mathbf{5.42\ mg/L.}$

Chapter 6

1. a) $O_2 = (7.2 \times 10^{-7})(46.4^{3.2})(0.225^{-0.194}) = 0.207\ mg/100\ mg/day.$

$\dfrac{0.207\ mg}{100\ mg\ day} \times \dfrac{10^4\,(100\ mg)}{kg} \times \dfrac{day}{24\ hr} = \mathbf{86.2\ mg/kg\ hr.}$

b) $O_2 = (1.9 \times 10^{-6})(46.4^{3.13})(0.26^{-0.138}) = 0.376\ mg/100\ mg\ day$

$= \mathbf{156.7\ mg/kg\ hr.}$

c) $O_2 = (4.9 \times 10^{-5})(52^{2.12})(0.092^{-0.194}) = 0.338\ mg/100\ mg\ day$

$= \mathbf{140.8\ mg/kg\ hr.}$

d) $O_2 = (3.05 \times 10^{-4})(58^{1.855})(0.40^{-0.138}) = 0.646\ mg/100\ mg\ day$

$= \mathbf{269.2\ mg/kg\ hr.}$

2. b) OD = $(75)(117.9^{-0.196})(10^{0.055 \times 8})$ = **81.1 mg/kg hr.**
   d) OD = $(249)(181.4^{-0.142})(10^{0.024 \times 14.4})$ = **263.7 mg/kg hr.**

3. a) Oc = $\dfrac{3 \times 1.5 \times 0.361}{214.4}$ (9155.23) = **69.4 mg/kg hr.**

b) Oc = $\dfrac{3 \times 1.5 \times 0.361}{220}$ (9155.23) = **67.6 mg/kg hr.**

c) Oc = $\dfrac{3 \times 1.5 \times 0.566}{159.5}$ (9155.23) = **146.2 mg/kg hr.**

d) Oc = $\dfrac{3 \times 1.5 \times 0.786}{254}$ (9155.23) = **127.5 mg/kg hr.**

Note: Rainbow trout growth equations were used for salmon and trout.

4. DO available = 7.75 − 2.43 = 5.32 mg/L.
   Log O_2 = $-0.999 - 0.000957(50) + 0.0000006(50^2) + 0.0327(28)$
   $- 0.0000087(28^2) + 0.0000003(50 \times 28)$

$$O_2 = \frac{0.731 \text{ mg DO}}{\text{g hr}} \times 5000 \text{ g} \times \frac{\text{L}}{5.32 \text{ mg}} \times \frac{\text{ft}^3}{28.31 \text{ L}} \times \frac{1}{8 \text{ ft}^3} = \frac{3.03}{\text{hr}} = 0.33 \text{ hr}$$
$$= \textbf{20 min.}$$

$$\text{Oc} = \frac{3 \times 1.5 \times 1.49}{184.5} (9155.23) = 332.7 \text{ mg/kg hr} = \textbf{43 min.}$$

5. $\text{Oc} = \dfrac{3 \times 1.5 \times 1.41}{335.2} (9155.23) = 173.3 \text{ mg/kg hr.}$

$$\frac{173.3 \text{ mg}}{\text{kg hr}} \times \frac{\text{kg}}{2.2 \text{ lb}} \times 13227.5 \text{ lb} \times \frac{\text{lb}}{453,600 \text{ mg}} \times \frac{\text{HP hr}}{1 \text{ lb}} = \textbf{2.3 HP.}$$

$$\log O_2 = -0.364. \ O_2 = \frac{0.433 \text{ mg}}{\text{g hr}} \times \frac{453.6 \text{ g}}{\text{lb}} \times 13227.5 \text{ lb} \times \frac{\text{lb}}{453,600 \text{ mg}}$$

$$\times \frac{\text{HPh}}{1 \text{ lb}} = \textbf{5.7 HP.}$$

6. 6.5: Oc = **159.4 mg/kg hr.**
   6.6: $O_2 = 0.415$ mg/g hr = **415 mg/kg hr.**

7. $\dfrac{5}{10.50} \times 153.14 = 72.9 \text{ mm Hg} = \dfrac{\text{DO}}{7.81} \times 158.15. \ \textbf{DO = 3.6 mg/L.}$

8. Location 1: 75 mm Hg $= \dfrac{\text{DO}}{10.53} \times 155.4 = 5.08 \text{ mg/L.}$

   10.53 − 5.08 = **5.45 mg/L.**
   Location 2: 75 mm Hg $= \dfrac{\text{DO}}{7.35} \times 155.4 = 3.55 \text{ mg/L.}$

   7.35 − 3.55 = **3.80 mg/L.**

9. $\text{Oc} = \dfrac{3 \times 1.5 \times 0.640}{76.2} \times 9155.23 = \dfrac{346.0 \text{ mg}}{\text{kg hr}} \times 2454.5 \text{ kg fish} \times \dfrac{\text{gal}}{3.785 \text{ L}}$

   $\times \dfrac{\text{min}}{600 \text{ gal}} \times \dfrac{\text{hr}}{60 \text{ min}} = 6.23 \text{ mg/L used.}$

   Ce = 9.79. Effluent DO = 9.79 − 6.23 = **3.56 mg/L.**
   $\dfrac{3.56}{9.79} \times 150.15 = \textbf{54.6 mm Hg.}$

Chapter 7

1. Use 10°C, 10-in. trout and BP = 760 mm Hg.
   Equation 7.1:
   $$CC = \frac{(10.92 - 5.14)(0.0545)}{0.00815} = \textbf{38.65 lb/gpm} \text{ (first pond).}$$

   $$CC = \frac{(8.74 - 5.14)(0.0545)}{0.00815} = \textbf{24.07 lb/gpm} \text{ (second and third ponds).}$$

Equation 7.2:

   CC = 1.80 × 10 = **18 lb/gpm** (first pond).

   CC = 1.44 × 10 = **14.4 lb/gpm** (second and third ponds).

   (Refer to Figure 7.2.)

2. CC = $\dfrac{0.5(7.64 - 3.60)}{0.469}$ = 4.31 lb/gpm × 1000 cfs = **1.93 × 10⁶ lb.**

   Rearing volume required = **1.93 × 10⁵ ft³.**

3. Elevation is 2000 feet above sea level.

   CC = $\dfrac{(10.65 - 5.42)(0.0545)}{0.0096}$ = 29.69 lb / gpm × 3000 gpm × 4.88 fish / lb

   = **434,912 fish.**

   89,070 lb × ft³/10 ft × ft of length/12 ft³ = **742.25 ft.**

4. CC = $\dfrac{(8.19 - 4.65)(0.0545)}{0.0118}$ =16.35 lb/gpm × 450 gpm × 16 = **117,720 lb.**

5. CC = $\dfrac{0.5(7.33 - 3.45)}{0.894}$ = 2.17 lb/gpm = 0.986 kg/gpm, for 100,000 kg, need

   1.014 × 10⁶ gpm.

6. The graph should show carrying capacities using the four formulae on the vertical axis vs. temperature for a given size of fish or size for a given temperature on the horizontal axis.

7. Carrying capacities for 10-in. trout at 10°C and BP of 760 mm Hg

   Equation 7.1: CC = $\dfrac{(10.92 - 5.14)(0.0545)}{0.00815}$ = 38.65 lb/gpm,

   Density = 19.32 lb/ft³.

   Equation 7.2: CC = 1.80 × 10 = 18 lb/gpm, Density = **9 lb/ft³.**

   Equation 7.4: CC = $\dfrac{1.2(10.92 - 5.14)}{0.448}$ = 29.25 lb/gpm, Density = **14.62 lb/ft³.**

   Equation 7.5: CC = $\dfrac{499(10.92 - 5.14)}{96}$ = 30.04 lb/gpm, Density = **15 lb/ft³.**

   According to Equation 7.8, Density = 10 × 0.5 = **5 lb/ft³.**

8. Use Equation 7.4: CC = $\dfrac{1.2(10.79 - 5.08)}{0.333}$ = 20.58 lb/gpm × 450 gpm × 12.39

   fish/lb = **114,758 fish.**

9. C_{1500} = $\dfrac{1750}{1200}$ = 1.46.

   C_{1500} = $\dfrac{1750}{1200}$ = 1.17.

   At 10°C for 10-in. trout, F_{1200} = $\dfrac{3 \times 1.46 \times 0.019}{10}$ = 0.0083.

$$F_{1500} = \frac{3 \times 1.17 \times 0.019}{10} = 0.0067$$

$$CC_{1200} = \frac{(10.92 - 5.14)(0.0545)}{0.0083} = \textbf{37.95 lb/gpm.}$$

$$CC_{1500} = \frac{(10.92 - 5.14)(0.0545)}{0.0067} = \textbf{47.02 lb/gpm.}$$

10. 14-in. catfish weighs 0.789 lb, 2.79 fish/kg × 200 kg/m³ × 0.94 m³ = **524 fish.**

Chapter 8

1. $CC = \frac{(9.76 - 4.95)(0.0545)}{0.0106} = 24.53$ lb/gpm × 450 gpm × 2.5 fish/lb = **27,596 fish** in first pond.
Cb = 0.093 (9.76 − 4.95) + 4.95 = 5.40 mg/L.
$CC = \frac{(5.40 - 4.95)(0.0545)}{0.0106} = \times 450$ gpm × 2.5 fish/lb = **2603 fish** in second and third ponds separated by weirs.
Cb = 0.34 (9.76 − 4.95) + 4.95 = 6.58 mg/L.
$CC = \frac{(6.58 - 4.95)(0.0545)}{0.0106} = \times 450$ gpm × 2.5 fish/lb = **9459** fish in second and third ponds separated by lattices.
Difference = 5485 lb = **$8,227.00.**

2. Note: when fish in the downstream pond have grown to 6 in., fish in the upstream pond will have reached 9 in. in length.
$Lr = \frac{(9.76 - DO)(0.0545)}{0.0119} = \frac{5832\ lb}{450\ gpm}$, DO = 6.93 mg/L.

Cb = 0.301 (9.76 − 6.93) + 6.93 = 7.78 mg/L.
$CC = \frac{(7.78 - 4.95)(0.0545)}{0.0177} \times 450$ gpm × 11.57 fish/lb = **45,385 fish.**

3. Required aeration can be accomplished by gravity devices alone.
4. 2 mg/L × 28.31 L/ft × 11 ft/sec × lb/453,600 mg × HPh/0.6 lb × 3600 sec/hr = 8.24 HP = **8.5 HP.**
8.5 HP/0.9 × 0.745 kW/HP × 8760 hr/year × $0.12/kWh = **$7396.36.**

5. $RT = 2.5 \times \frac{(10.92 - 5)(1.025^{-10})(0.85)}{8.84} = 1.11$ lb/HPh.

$\frac{1.11\ lb}{HPh} \times \frac{2HP}{} \times \frac{453,600\ mg}{lb} \times \frac{gal}{3.785\ L} \times \frac{min}{3000\ gal} \times \frac{hour}{60\ min}$

= 1.48 mg/L + 5 = **6.48 mg/L.**

Second aerator adds 1.11 mg/L, final DO = **7.59 mg/L.**
6. Ce = 9.41 mg/L, 90% of Ce = 8.47 mg/L.
$\frac{1\ lb}{HPh} \times \frac{453,600\ mg}{lb} \times \frac{L}{3.47\ mg} \times \frac{gal}{3.785\ L} \times \frac{min}{1000\ gal} \times \frac{hr}{60\ min}$

= 1.735 HP = **1.75 HP.**

$$CC = \frac{(8.47 - 5)(0.0545)}{0.0122} = 15.54 \times 1000 \times \$0.25 = \textbf{\$3885.20 benefit.}$$

11-in. trout takes 12.7 months to grow.

$$\frac{1.75\,HP}{0.9} \times \frac{0.745\,kW}{HP} \times \frac{9144\,hr}{12.7\,month} \times \frac{\$0.11}{kWh} = \$1465.55 \text{ electric}$$

$$+\$1200.00 \text{ ownership cost} = \textbf{\$2665.55.}$$

7. Each nozzle delivers 125 gpm so 40 nozzles are required. Each nozzle sprays into a 2-ft^2 area of basin so the surface area of the basin is 80 ft^2.

8. Ce = 10.43 mg/L, Ce'$_t$ = 31.29 mg/L.

$$RT = 1.5 \times \frac{(31.29 - 5)(1.025^{-8})(0.85)}{8.84} = 3.11\,kg/kWh.$$

$$\frac{3.11\,kg}{kWh} \times \frac{0.745\,kW}{HP} \times \frac{2\,HP}{} \times \frac{10^6\,mg}{kg} \times \frac{gal}{3.785\,L} \times \frac{min}{1000\,gal}$$

$$\times \frac{hr}{60\,min} = 20.4\,mg/L + 5 = 25.4\,mg/L.$$

The compressor is too large because the calculations indicate a supersaturated effluent. The desired effluent DO is 9.39 mg/L so a compressor designed to deliver 4.39 mg/L is required.

$$\frac{4.39\,mg}{L} \times \frac{kg}{1,000,000\,mg} \times \frac{kWh}{3.11\,kg} \times \frac{HP}{0.745\,kW} \times \frac{3.785\,L}{gal} \times \frac{1000\,gal}{min}$$

$$\times \frac{60\,min}{hr} = 0.43\,HP = \textbf{0.5 HP.}$$

9. $$\frac{4.39\,mg}{L} \times \frac{3.785\,L}{gal} \times \frac{1000\,gal}{min} \times \frac{1440\,min}{day} \times \frac{lb}{453,600\,mg} \times \frac{1}{0.9}$$

$$\times \frac{\$0.07}{lb} = \$4.10/day \text{ for oxygen.}$$

$$\frac{0.75\,HP}{0.9} \times \frac{0.745\,kW}{HP} \times \frac{24\,hr}{day} \times \frac{\$0.10}{kWh} = \textbf{\$1.49/day for compressor.}$$

10. Ce$_{DO}$ = 10.9, DO = 2.72, desired DO = 9.81.
    Ce$_{DN}$ = 18.54, DN = 23.17, desired DN = 17.09.
    Say RS = 2 lb/HPh,

$$RT_{DO} = 2\frac{(10.9 - 2.72)(1.025^{-11})(0.85)}{8.84} = 1.2\,lb/HPh.$$

$$\frac{7.09\,mg}{L} \times \frac{3.785\,L}{gal} \times \frac{500\,gal}{min} \times \frac{60\,min}{hr} \times \frac{lb}{453,600\,mg} \times \frac{HPh}{1.2\,lb} = 1.48\,HP.$$

$$RT_{DN} = 1.51 \times 2\frac{(23.17 - 18.54)(1.025^{-11})(0.85)}{14.88} = 0.61\,lb/HPh.$$

$$\frac{6.08\,mg}{L} \times \frac{3.785\,L}{gal} \times \frac{500\,gal}{min} \times \frac{60\,min}{hr} \times \frac{lb}{453,600\,mg} \times \frac{HPh}{0.61\,lb} = \textbf{2.60 HP.}$$

To obtain satisfactory levels of both gasses, a 2.6-HP unit is required.

Chapter 9

1. $A = 56P = 56(0.25) = \dfrac{14 \text{ g TAN}}{\text{kg food day}} \times 54.4 \text{ kg food} \times \dfrac{1000 \text{ mg}}{\text{g}} = 763{,}000 \text{ mg}$

    TAN produced per day.

    $763{,}000 \text{ mg} \times \dfrac{\text{ft}^3}{28.31 \text{ L}} \times \dfrac{1}{5 \text{ ft deep}} = 0.124 \text{ mg}/\text{L added} + 1.0 = \mathbf{1.124 \text{ mg/L.}}$

2. $f_{6.5} = \dfrac{1}{10^{9.155-6.5} + 1} = 0.0022.$ $\text{TAN} = \dfrac{\text{NH}^3}{f} = \dfrac{0.02}{0.0022} = \mathbf{0.91 \text{ mg/L.}}$

    $f_{9.5} = \dfrac{1}{10^{9.155-9.5} + 1} = 0.689.$ $\text{NH}_3 = (\text{TAN})(f) = \mathbf{0.63 \text{ mg/L.}}$

3. $A = 56(0.45) = 25.2$ g TAN/kg food $= 11.45$ g TAN/lb food.

    $F = \dfrac{3 \times 1.5 \times 0.021}{12} = 0.0078 \text{ lb food}/\text{lb fish.}$

    $f = 0.0096.$ $\text{TAN} = \dfrac{0.013}{0.0096} = 1.35 \text{ mg}/\text{L} = \text{maximum TAN.}$

    $\dfrac{\text{lb fish day}}{0.0078 \text{ lb food}} \times \dfrac{\text{lb food}}{11.45 \text{ g TAN}} \times \dfrac{\text{g}}{1000 \text{ mg}} \times \dfrac{1.35 \text{ mg}}{\text{L}} \times \dfrac{3.785 \text{ L}}{\text{gal}} \times \dfrac{100 \text{ gal}}{\text{min}}$

    $\times \dfrac{1440 \text{ min}}{\text{day}} = \mathbf{8239 \text{ lb fish.}}$

4. One complete water use for reaerated water (influent DO at 90% saturation)

    $= \dfrac{(9.73 - 5.22)(0.0545)}{0.015} = 16.13 \text{ lb/gpm} \times 3 \text{ uses} = 48.4 \text{ lb/gpm.}$

    $A = \dfrac{56(0.48)}{2.2} = \dfrac{12.21 \text{ g TAN}}{\text{lb food}} \times \dfrac{0.015 \text{ lb food}}{\text{lb fish}} \times \dfrac{48.4 \text{ lb fish min}}{\text{gal}} \times \dfrac{\text{gal}}{3.785 \text{ L}}$

    $\times \dfrac{\text{day}}{1440 \text{ min}} \times \dfrac{1000 \text{ mg}}{\text{g}} = 1.63 \text{ mg/L TAN after 3 water uses.}$

    $f = 0.0077$ when $\text{NH}_3 = 0.0125.$ $0.0077 = \dfrac{1}{10^{9.75-\text{pH}} + 1}$

    Solve for pH. $\mathbf{pH = 7.64.}$

5. $\text{CC} = \dfrac{(6.27 - 3.55)(0.0545)}{0.0179} = 8.26 \text{ lb/gpm.}$

    $A = \dfrac{56(0.32)}{2.2} = 8.14 \text{ g TAN}/\text{lb food.}$

$$TAN = \frac{8.14 \text{ g}}{\text{lb food}} \times \frac{0.0179 \text{ lb food}}{\text{lb fish}} \times \frac{8.26 \text{ lb fish min}}{\text{gal}} \times \frac{\text{day}}{1440 \text{ min}}$$

$$\times \frac{1000 \text{ mg}}{\text{g}} = 0.22 \text{ mg} / \text{L}.$$

f = 0.717 so NH_3 = 0.158 mg/L. If this is lost to the atmosphere, 0.62 mg/L TAN will remain and the influent NH_3 concentration to the second set of ponds will be **0.044 mg/L.**

6. CC = $\dfrac{(8.23 - 4.66)(0.0545)}{0.015}$ = 12.9 lb/gpm $\times$ 10 = 129.2 lb/gpm.

$$\frac{12.22 \text{ g TAN}}{\text{lb food}} \times \frac{0.015 \text{ lb food}}{\text{lb fish}} \times \frac{129.2 \text{ lb fish min}}{\text{gal}} \times \frac{\text{gal}}{3.785 \text{ L}} \times \frac{1000 \text{ mg}}{\text{g}}$$

$$\times \frac{\text{day}}{1440 \text{ min}} = 4.35 \text{ mg} / \text{L TAN}.$$

f = 0.0016 $\times$ 4.35 = **0.0072 mg/L NH_3.** Ammonia filters are not required because 0.0072 mg/L is less than reported maximum safe levels for NH_3.

7. $\dfrac{176 \text{ kg}}{\text{m}^3} \times 67.5 \text{ m}^3 \times \dfrac{0.01 \text{ kg food}}{\text{kg fish}} \times \dfrac{25.2 \text{ g TAN}}{\text{kg food}} \times \dfrac{\text{sec}}{113 \text{ L}} \times \dfrac{1000 \text{ mg}}{\text{g}}$

$$\times \frac{\text{day}}{86,400 \text{ sec}} \times 0.0058 \text{ (f)} = \textbf{0.0018 mg/L } \mathbf{NH_3}.$$

8. Compile the following table tabulating cumulative oxygen consumption and specific growth rate.

Serial unit	Cumulative oxygen consumption	Specific growth rate
1	2.5	0.643
2	5.1	0.667
3	7.5	0.500
4	9.9	0.310
5	12.4	0.143

Regress specific growth rate (SGR) against cumulative oxygen consumption (COC) or draw a graph with SGR on the horizontal axis and COC on the vertical axis. The coefficient of correlation is –0.954, the slope is –16.6, and the intercept is 15.0. The equation for the line is COC = 15.0 – 16.6(SGR). The SGR in the first container is 0.643 and half this rate is 0.321. Thus, $ECOC_{50}$ = 15.0 – 16.6(0.321) = **9.67 mg/L.**

To estimate carrying capacity, regress COC against cumulative fish load using the following data.

Serial unit	Cumulative fish load	Cumulative oxygen consumption
1	18.7	2.5
2	37.8	5.1
3	54.0	7.5
4	67.6	9.9
5	79.2	12.4

The resulting regression equation is Fish Load = 5.58 + 6.13 (COC) and the carrying capacity is the fish load resulting from a COC equal to the $ECOC_{50}$: CC = 5.58 + 6.13(9.67) = 64.8 lb/2 gpm = **32.40 lb/gpm.**
Carrying capacity with respect to oxygen is

$$CC = \frac{[(0.9 \times 10.64) - (5.01)][0.0545]}{0.033} = 7.54 \text{ lb/gpm.}$$

Metabolite considerations allow for $\dfrac{32.40}{7.54} = 4.3$ water uses so the hatchery

for which the PCA was conducted should be designed with rearing units placed **four in series.** Note: The above calculation was based on 52° F water and 3-in. fish.

9. $\dfrac{0.033 \text{ lb food}}{\text{lb fish}} \times \dfrac{11.45 \text{ g TAN}}{\text{lb food}} \times 0.00636 \times \dfrac{1000 \text{ mg}}{\text{g}} = \dfrac{2.4 \text{ mg NH}_3}{\text{lb fish day}}.$

$\dfrac{\text{lb fish day}}{2.4 \text{ mg}} \times \dfrac{0.016 \text{ mg NH}_3}{\text{L}} \times \dfrac{3.785\text{L}}{\text{gal}} \times \dfrac{1440 \text{ min}}{\text{day}} = \dfrac{\textbf{36.34 lb}}{\textbf{gpm}}.$

10. Fish should reach 14 in. and be removing 30% of the influent DO (90% of saturation) at the end of the 6-week test. ΔL = 1.49 mm (0.059 in.)/day so the bioassay units should be stocked with **11.5-in. fish.**

$O_2 = \dfrac{0.444 \text{ mg}}{\text{g hr}} \times \dfrac{(6.97 \times 0.3) \text{ mg}}{\text{L}} \times \dfrac{\text{min}}{20 \text{ L}} \times \dfrac{\text{hr}}{60 \text{ min}} = 5649 \text{ g fish.}$

14-in. catfish weigh 358 g each so **15 fish** should be stocked per bioassay unit.

Chapter 10

1. The following equations are required: tilapia regression equation from Table 2.10 and Equations 2.1, 2.2, 9.1, and 10.1. Value used for C in 2.2 was 1.7 and 0.35 was used for P in 9.1. The required surface is 5468 ft². **A biodisk containing 387 3-ft diameter disks would fit into a 16-ft long trough if the disks were 0.25 in. apart.**

2. The required system flow is 406 gpm from 7.1 when Oa and Ob are 7.53 and 3.55 mg/L. **The filter must have a cross-sectional area of 162 ft² and contain 137 ft³ of media (Table 10.1). Media depth would then be slightly greater than 10 in.**

3. 75 mm Hg = 3.55 mg/L. 784 g TAN × 4.33 g DO/g TAN = 3394.72 g DO/day. Divide by the system flow to get DO drop of 1.54 mg/L across the biofilter. **Effluent DO is 2.01 mg/L so aeration is not required.**

4. 784 g TAN × 7.14 g $CaCO_3$/g TAN = 5598 g $CaCO_3$ per day. To correct for $NaHCO_3$ multiply by the ratio of the formula weights, 84/100. The daily requirement of sodium bicarbonate is **4702 g.**

5. Three 42-in. 5-HP sand filters would provide a minimum flow of 393 gpm. **At least 203 ft^2 of tube settler top surface area is required.**

6. 0.25 mg/L N × 46/14 = 0.82 mg/L NO_2^- × 6 = 4.93 mg/L Cl^- × 58/35 = **8.17 mg/L NaCl.**

7. The biofilter DO requirement is 3.39 kg/day (Problem 3). The fish require 8.80 kg/day (Equation 7.1). The monthly requirement is 366/0.90 = **407 kg.**

8. When C = 1.7, carrying capacity = 14.75 kg/gpm. When P = 0.45, 90% reuse results in a TAN concentration of 5.2 mg/L. If the maximum allowable NH_3 concentration is 0.016 mg/L, **the pH must be 7.04 or less.**

9. When C = 1.7, P = 0.45 and maximum NH_3 = 0.016 mg/L, **498 gpm** of the system flow must be new water and the hatchery operates on 80% reuse without biofilters.

10. **12,727 lb of fish** due to increased rate of nitrification at the higher temperature.

UNIT ABBREVIATIONS USED

Atmosphere	atm
Calorie	C
Centimeter	cm
Cubic feet per second	cfs
Degrees Celsius	°C
Degrees Fahrenheit	°F
Dollar	$
Foot	ft
Gallon	gal
Gallon per minute	gpm
Gram	g
Horsepower	HP
Horsepower hour	HPh
Hour	hr
Inch	in.
Inches of mercury	in. Hg
Kilogram	kg
Kilowatt	kW
Kilowatt Hour	kWh
Liter	L
Meter	m
Micrometer	μm
Milliequivalents per liter	meq/L
Milligram	mg
Millimeters of mercury	mm Hg
Monthly Temperature Unit	MTU

Pound	lb
Pounds per square inch	psi
Revolutions per minute	rpm
Second	sec
Temperature Unit	TU

ESSENTIAL UNIT CONVERSION FACTORS

1 kg = 1000 g
1 g = 1000 mg
1 kg = 2.2 lb
1 kg = 10^6 mg
1 ton = 2000 lb
1 lb = 453.6 g
1 lb = 453,600 mg
1 m = 100 cm
1 cm = 10 mm
1 m = 3.28 ft
1 ft = 12 in.
1 mile = 5280 ft
1 in. = 2.54 cm
1 m^3 = 1000 L
1 ft^3 = 7.48 gal
1 acre = 43,560 ft^2
1 gal = 3.78 L
1 cfs = 448.8 gpm
1 gal of water weighs 8.34 lb @ 4°C
1 gal of water weighs 8.32 lb @ 25°C
1 L of water weighs 1 kg @ 4°C
1 year = 365 days
1 day = 24 hr
1 hr = 60 min
1 min = 60 sec
1 HP = 0.746 kW
1 HP = 33,000 ft lb/min
1 HP = 2545 BTU
1 atm = 760 mm Hg
1 mm Hg = 0.0193 psi
°C = 5/9 (°F − 32)
°F = 9/5 °C + 32

Index

Milton Keynes UK
Ingram Content Group UK Ltd.
UKHW040052071024
449327UK00019B/493